Beneficial Nematodes and Nematode Antagonistic Bioagents

NIPA® GENX ELECTRONIC RESOURCES & SOLUTIONS P. LTD.
New Delhi-110 034

Beneficial Nematodes and Nematode Antagonistic Bioagents

Edited by

S. Subramanian
P. Vetrivelkalai
M. Sivakumar
K. Poornima

Department of Nematology
Tamil Nadu Agricultural University
Coimbatore - 641 003, India

NIPA® GENX ELECTRONIC RESOURCES & SOLUTIONS P. LTD.
New Delhi-110 034

NIPA® GENX ELECTRONIC RESOURCES & SOLUTIONS P. LTD.

101,103, Vikas Surya Plaza, CU Block
L.S.C. Market, Pitam Pura, New Delhi-110 034
Ph : +91 11 27341616, 27341717, 27341718
E-mail: newindiapublishingagency@gmail.com
www: www.nipabooks.com

For customer assistance, please contact
Phone: + 91-11-27 34 17 17
Fax: + 91-11- 27 34 16 16
E-Mail: feedbacks@nipabooks.com

ISBN: 978-81-19002-72-6

Composed and Designed by NIPA®.

Preface

Nematology is an emerging science deals with nematodes infesting crop plants, free living and beneficial nematodes. Plant parasitic nematodes are biotrophic parasites, more than 4100 species associated with crop plants, which cause 12.3 per cent economic yield loss with the tune of $ 157 billion in worldwide. Many strategies have been developed for nematode management including the use of chemical nematicides, which are harmful to human health and environment. Biological control is now accepted as an alternative, economic and environmentally sound management strategy. The beneficial microbe's *viz*., bacteria, fungi and actinomycetes play a main role in stimulating defence mechanism and promote the plant growth and also showed nematicidal activities by production of secondary metabolites and toxic compounds.

On the other hand, a large number of beneficial nematodes especially entomopathogenic nematodes like *Steinernema* and *Heterorhabditis* having mutualistic bacteria *Xenorhabdus* and *Photorhabdus,* respectively, cause rapid death of insect host within short period. Entomopathogenic nematodes have been used as potential biocontrol agents against many insect pests *viz*., Lepidoptera, Coleoptera, Hemiptera and Dictyoptera in agri-horticultural crops.

Based on the biocontrol potential of entomopathogenic nematodes it can be used in the pest combat strategies and bacteria, fungi and actinomycetes having biocontrol potential against plant parasitic nematodes. Keeping this view a book on "*Beneficial Nematodes and Nematode Antagonistic Bioagents*" has been formulated. This book comprises of two sections, the first section (chapter 1 to 16) deals with isolation, identification, molecular characterization, mass multiplication, formulations and shelf life of bacteria, fungi and actinomycetes. The second section (chapter 17 to 24) deals with isolation, identification, mass culturing of insect host and mass multiplication, formulations and field potential of entomopathogenic nematodes against economically important insect pests.

The help rendered by Dr. R.K Walia, Dr. A.S. Krishnamoorthy, Dr. K. Eraivan, Dr. M.Muthamilan, Dr. S. Nakkeeran, Dr. S. Sridharan, Dr. C. Sankaranarayanan, Dr.J. Gulsar Banu, Dr. A. Shanthi and Dr. N. Swarnakumari in assisting the technical corrections are greatly acknowledged. We gratefully acknowledges the scientists and research scholars of Department of Nematology, Plant Pathology and Agricultural Entomology, Tamil Nadu Agricultural University, Coimbatore for having contributed chapters in this book.

Editors

Contents

PART I
Bioagents for Plant Parasitic Nematodes

1

Biological Control – Revisiting the Basic Principles

R.K. Walia

Project Coordinator, AICRP (Nematodes), ICAR-IARI, New Delhi- 110012

The basic principles of biological control of pests and diseases remain the same, whether applied to insect pests, nematodes or plant pathogens. In the recent years, there is a paradigm shift towards studying antibiosis and its application in biological suppression of pathogens, yet the classical bio-control will remain as relevant as ever. The following discussion is based on recapitulating the basic principles of classical biological control as applicable to plant parasitic nematodes.

Biological control agents (BCAs) have been exploited for pest control in different ways. One way is to manipulate the naturally occurring BCAs by augmenting their populations to pest suppressive levels. In this approach no BCAs are introduced, but various means such as agronomic practices, introduction of organic matter etc. are adopted to favour the build-up of BCAs. The second way is the introduction of BCAs, either as inoculative or inundative release. In the former case, the BCAs are introduced in small quantities and allowed to build-up gradually in due course of time. However, in the latter approach, BCAs are introduced in large quantities so that quick pest suppression can be achieved. Each method has its own merits and demerits (Table 1).

Table 1: Useful characteristics of an organism required for three types of biological control

Characteristic	Inundative	Inoculative	Natural
Rapid colonization in soil	-	+	-
Persistence	-	+	+
Virulence	+	+	+
Predictive control below economic threshold	+	+	+
Easy production and application	+	+	-
Good storage	+	+	-
Low cost	+	+	-
Safe	+	+	+
Compatibility with agrochemicals and farm practices	+	+	+

The Ecological Basis of Biological Control

The biotic environment of plant consists of the association of different interdependent species in nature. Individuals of a particular species form populations which in association with other species constitute communities. Within a community we can recognize primary produces, primary consumers, secondary consumers, decomposers and scavengers, thus forming food chains. In the context of biological control, the organisms which live at the expense of other organisms are described as natural enemies. And within a community there is hardly any species which is free from such natural enemies. This phenomenon is, however, dynamic and is influenced by several other factors. In nature, the growth of pest populations may be checked by limited resources (food, space, shelter); periodically occurring inclement weather (heat, draught, rain etc.); competition among themselves or other kinds of animals; and natural enemies (parasites, pathogens, predators). While fluctuations of resources, weather and biological competitions are temporary phenomena, the natural enemies are almost universally present, and often significantly so. The forces that regulate natural populations of organisms may be density dependent or density independent. In the context of natural biological control, the occurrence of diseases (among density dependent forces) is key to the long-term suppression of pests, while polyphagous parasites and predators (among density independent forces) enforce only temporary suppression of pests.

Nematode Ecosystem

The following description of nematode ecosystem is a reproduction of my earlier publication "*Imagine a cubic meter of piece of field soil with its millions of soil particles, big and small, enclosing between them a network of labyrinthine pores and spaces filled with air and moisture. Consider the flooded situation when the air is replaced with water, soil particles tend to come closer, the soil chemicals get diluted and again become concentrated when water evaporates. Imagine this physical and chemical structure coming to life with the teeming millions of micro-organisms – protozoans, bacteria, fungi, viruses and a few macro-organisms like insects etc. Perceive the growth of plant roots within this system – displacing soil particles as they proliferate, respiring, secreting and exuding gases and chemicals. Interspersed in this piece of soil are thousands of nematodes of different kinds – the free living (microbivorous), the predators, and a few hundreds of them plant parasites*".

Arable soils are a biotic complex. It has been estimated that one cubic meter of highly fertile soil may contain as many as 300g bacterial cells, 39 g protozoa, 12 g nematodes, and 400 g fungi, besides other organisms. It is logical to understand, therefore, that biotic interactions are natural phenomena; and that the biological activities of each component of this ecosystem are directly or indirectly influenced by the concomitant components. It is unimaginable to perceive the complexity and magnitude of this ecological web. However, taking into account the plant parasitic nematodes at the center stage, two important aspects of interactions emerge, (i) the pathological capabilities of phytonematodes as influenced by other organisms or vice versa (role of nematodes in disease complexes);and (ii) the antagonistic effect of other organisms on the biological activities of phytonematodes (biological control of phytonematodes).Nematode antagonists belong to diverse groups: these include fungi, bacteria, viruses, collemboles, tardigrades, enchytreids, predacious nematodes, mites etc.

Plant species and environmental conditions greatly affect the quality and quantity of nutrients released into the rhizosphere and the structure of the microbial community around roots, but the influence of these factors on the microbial natural enemies of nematodes has been little studied. Between 10-30% of the carbon assimilated by the plant is released into the rhizosphere and this supports microbial activity. Observations on different plants in different soils indicate that fungi are mostly in the vegetative state in the rhizosphere, whereas in the bulk soil, 70-90% of the propagules are spores. Despite the large microbial populations in the rhizosphere, bacteria occupy <10% of the root surface and that fungal hyphal densities are only 12-14 mm m^{-2} root. At such densities, mobile nematodes may readily avoid their natural enemies and select uncolonised sections of root on which to feed.

The foregoing description of plant environment or in relation to nematodes – the soil ecosystem – opens up a Pandora's Box. Biological control, in essence, is an ecological problem. Hence an understanding of the ecological basis is essential for developing a successful biological control programme.

Development of Biocontrol Agent of Phytonematodes

The development of BCAs involves three steps.

1. Isolation and Characterization

Search avenues for BCAs: The development of nematodes and their antagonists co-evolves. In the areas where nematodes have been newly introduced, their population grows exponentially in the initial years and thereafter it tends to stabilize. The natural enemies play a vital role in population regulation. Hence, the area to be surveyed for nematode antagonists should have a long history of nematode infestation, as such sites are likely to yield a wider spectrum of nematode enemies. Nematode antagonists also fluctuate seasonally; therefore, repeated sampling (surveillance) may also be required.

Diversity of BCAs: Since hosts and parasites co-evolve, it is also an established fact that greater genetic diversity of any plant parasitic nematode species would occur in the geographic areas where its host crop originated. As for golden nematode of potato, Andes mountains (Peru) are considered to be the origin of potato, therefore, it would reveal maximum diversity in golden nematode species complex. And for the same reason, the BCAs of this nematode are most likely to evolve and flourish in such areas.

Indigenous vs Exotic strains: Indigenous organisms for bio-control are more adapted to the local agro-climatic conditions, and this is of vital importance in the development of any BCA. Use of exotic organisms should be resorted to if the attempts on indigenous organisms are unsuccessful.

Temporal considerations: Time of sampling is another factor that can impact the results, and should be carefully planned on the basis of nematode biology, nature of parasitism (ectoparasite, endoparasite). For example, a search for the antagonists of *Heterodera avenae* would necessitate sampling during January/February if parasites of females are to be recorded; during September/ October for egg parasites (cysts remain in soil from April to November); and during November/December for larval parasites as they remain in soil for a very short period before invading the roots.

Choice of isolation techniques from soil: The methods based on the principle of nematode mobility are likely to yield fewer incidences of BCAs since a vast majority of the diseased nematodes are moribund and are rendered immobile or are killed. Such nematodes are not extracted by these techniques. Sugar floatation and centrifugal techniques and their modifications are best suited for this purpose and provide a true picture of the incidence of BCAs.

Techniques for the isolation of predacious fungi directly from soil have been the subject of much study. These include – soil sprinkling on agar plates, pouring of soil extract on agar plates, preferably with nematode baits. Endoparasitic fungi appear late in succession of appearance of nematode-destroying fungi by these techniques.

Isolation of fungi from infected nematodes (vermiform stages, eggs etc.) has not been standardized. Generally, nutrient deficient media are used initially, and gradual transfers and purifications through increase in the availability of nutrients yield most of the organisms in pure cultures. However, the pathogenic capability of the isolated organisms should invariably be confirmed by *in vitro* tests. It is not unusual to get saprophytic types or incidence of multiple infections in many cases. Endoparasites are generally slower to grow, and are usually masked by the fast growing saprophytes. The major problem is encountered with obligate parasites which do not grow at all on synthetic or artificial media, and therefore, would require special techniques involving host stages for their growth and subsequent identification.

2. Screening and Selection

The successful development of BCAs is dependent upon careful selection of suitable organisms. Such selection requires a detailed understanding of the epidemiology and mode of action of the agent to ensure that it is appropriate for the target nematode pest.

In vitro screening: Screening methods in nematode biocontrol are empirical and selection of effective agents is often based on intuition rather than experimental evidences. Screening involves testing the organism in laboratory for the pathogenicity/efficacy, as well as pot trials. Both are beset with inherent lacunae, compounded by the erroneous experimentation. An ill-designed experiment may lead to exclusion of a very promising agent.

Lab screening should form a part of the routine and should invariably precede pot experimentation. The isolated organism or the diseased nematode stage is kept on water agar, and observations are recorded on qualitative or quantitative capability of an organism to infect healthy nematodes. The nematode stage included in the test and the temperature play important role in inferences. The organisms which do not infect or infection capacity is too little be rejected at this stage, for it reduces the quantum of more cumbersome pot screening. This is more important in cases involving testing of several isolates of the same organism. Isolates vary in their virulence and less virulent strains can be eliminated at this stage.

Method of introduction: Several factors influence the efficacy of an organism in pot trials, and most important is the method of introduction of the organism. In a review of pot experiments involving *Purpureocillium lilacinum*, Kerry (1990) pointed out that more than 50% of the experiments were inappropriately designed. The method of introduction varies with the type of the agent, i.e., obligatory or facultative, and it is worthwhile to discuss merits and demerits of both at this stage (Table 2).

Table 2. Some important characteristics of obligate and facultative parasites of nematodes

Characteristic	Obligate	Facultative
Host range	Narrow	Wide
Resting structure	Usually present	Often absent
In vitro culture	Difficult	Easy
Food sources required to establish	No	Usually
Spread in soil	Passive	Active
Efficacy dependent on host density	Usually	Less so

Adapted from Kerry 1990

Green-house trials- Some observations: Both types of agents could be considered for the development of BCAs. In pot trials, often autoclaved soils are used which are nematised subsequently to obtain desired initial population. This one step rules out the competitive potential of the test organism with the resident microflora and fauna in field soils. Thus the results achieved may be an exaggeration of what may be expected in field trials. In case of facultative organisms, the addition of the food base acts as soil amendment which may itself result in nematode suppression. In pot experiments, recording observations on the establishment and survival of the introduced organism in autoclaved soil would be of little use since the competing organisms were eliminated. These experiments, however, could form an intermediary step to assess the potential of an organism in checking a multiplying nematode population on its host plant under controlled temperature etc., and should preferably be terminated after completion of one or two generations. The true potential of any organism would be discernible only under field conditions/micro-plotexperiments where nematode populations have been more or less uniformly maintained by growing susceptible hosts.

Establishment of BCAs: BCAs should not be construed as magic bullets that will eliminate nematode pests in one go. Conversely, the effect of any introduced BCA may not manifest itself in the first one or two crops. Hence its efficacy in terms of nematode population reduction, crop growth and yield, and more importantly its establishment and fluctuation must be monitored till 2-3 crops in order to gauge the true potential of the organism and its final selection. The establishment of organism's survivaland population assessment also creates added technical problems. Development of specific media in case of facultative parasites – *P. lilacinum* and *Pochonia chlamydosporia* prove handy; while in case of obligate parasites, other useful traits like – survival stage in soil or infected stages may be used quantitatively for this purpose.

Test Plant host: Selection of nematode host plant in such tests can be of significance, e.g., in case of root-knot nematodes and organisms involving egg parasites, in certain hosts a significant portionof the eggs may be retained within thehost tissues, while in other hosts, eggs may be exclusively laid on the root surface.

Obligate vs facultative BCAs: It is debatable whether obligate or facultative parasites would prove better BCAs. Both have been in use in other related sciences. Another point of discussion is the selection of juvenile parasites or those which parasitize females and eggs; perhaps the latter category finds more favour.

The results achieved through laboratory – green house – field trials reveal the potential of an organism, yet the final selection of any organism for development as a BCA should be based on a strong back-up information on its mode of action, epidemiology aspects and other safety parameters.

3. Mass Rearing and Field Testing

Some workers have attempted to mass multiply various microbial antagonists of nematodes for field testing; however, mostly actively growing fungi with large quantities of organic matter have been used. However, for field introductions, the formulations of microbial nematode antagonists would be in a different form, and are intended to be used in much lower quantities. Treatment with a BCA in excess of 200 kg per ha should be avoided to keep costs of production, storage and application at an economic level. Low costs, efficacy, compatibility with existing farm practices, and safety are important in determining the acceptance of biological control products by the growers.

To Sum Up

Undoubtedly, an ideal bio-control system should be stable, permanent and self-sustaining. BCAs of phytonematodes are abundant in nature; and an intensive search would reveal many potential candidate organisms which could be put to rigorous evaluation before making selections for their development as BCAs. Detailed information on the mode of action, survival and epidemiology should be generated with respect to promising organisms prior to their consideration for commercial development. Most work on BCAs of nematodes has involved root-knot and cyst nematodes only. Understandably, this may be linked to their economic importance and the ease with which these can be handled and noticed.

It should be emphatically realized that the nematologists working with BCAs are placed in a piquant situation compared to their colleagues dealing with nematicides. Those working on biocontrol have to search, study and then develop and formulate an "organism", which may require several years of intensive team approach. Secondly, the immediate demonstrable effect is usually not forthcoming in case of BCAs. The efficacy of the organisms needs to be monitored over longer times. This necessitates the blending of BCAs with other methods of nematode control; and to develop them as a component of integrated nematode management. Considering the long-term and ecological advantages of using BCAs, INM approaches based on a blend of BCAs are being advocated.

Suggested Readings

Kerry, B.R. 1990. An assessment of progress toward microbial control of plant parasitic nematodes. *Supplement of Journal of Nematology*, 22: 621-631.

Kerry, B.R. 2000. Rhizosphere interactions and the exploitation of microbial agents for the biological control of plant-parasitic nematodes. *Annual Review of Phytopathology*, 38: 423-441.

2

Bacterial Antagonists Against Phytonematodes

S. Nakkeeran and C. Dheepa

Department of Plant Pathology, Tamil Nadu Agricultural University Coimbatore- 641 003, Tamil Nadu

Introduction

The soil around plant roots that forms the rhizosphere is a dynamic, complex zone. All plant parasitic nematodes are obligate parasities and must enter this zone to reach their host and cause damage (Kerry, 2000). Nematodes in soil are subject to infections by bacteria and fungi. This creates the possibility of using soil microorganisms to control plant-parasitic nematodes. Bacteria are numerically the most abundant organisms in soil, and some of them, for example members of the genera *Pasteuria*, *Pseudomonas* and *Bacillus* have shown great potential for the biological control of nematodes. Nematophagous bacteria are distributed broadly, possess diverse modes of action, and have broad host ranges. A variety of nematophagous bacterial groups have been isolated from soil, host-plant tissues, and nematodes and their eggs and cysts (Siddiqui and Mahmood, 1999; Kerry, 2000; Meyer, 2003).

They affect nematodes by a variety of modes: for example parasitizing; producing toxins, antibiotics, or enzymes; interfering with nematode–plant-host recognition; competing for nutrients; inducing systemic resistance of plants; and promoting plant health (Siddiqui and Mahmood, 1999). They form a network with complex interactions among bacteria, nematodes, plants and the environment to control populations of plant-parasitic nematodes in natural conditions (Kerry, 2000).

Nematophagous Bacteria and their Modes of Action Against Nematodes

Parasitic Bacteria

Members of the genus *Pasteuria* are obligate, mycelial, endospore-forming bacterial parasites of plant-parasitic nematodes and water fleas. A number of bacterial species in this genus have shown great potential as biocontrol agents against plant parasitic nematodes. They occur worldwide and have been reported from at least 51 countries (Siddiqui and Mahmood, 1999). Members of the genus have been reported to infect 323 nematode species belonging to 116 genera, including both plant-parasitic nematodes and free-living nematodes. The majority of economically important plant-parasitic nematodes have been observed to be parasitized. However, coarse soil was found better for infection than fine clay particles. Excess watering had no adverse effect on the parasitic activity of *P.penetrans.* Obligate bacterial parasite, *Pasteuria penetrans*. The name *P. penetrans* is given to a group of spore forming bacteria which are parasitic to a number of important plant parasitic nematodes (Birchfield and Antonpoulos, 1976; Starr and Sayre, 1988). It is one of the most efficient natural enemies of root-knot nematodes (Gowen and Ahmed, 1990). The bacteria have been reported from nearly 200 nematode species from a wide range of environments (Sayre and Starr, 1985), but its occurrence and abundance seems to be variable. This variability is thought to be due to several factors, including differences in the specificity of isolates of *P. penetrans* to populations and species of *Meloidogyne* (Davies and Danks, 1993).

P. penetrans is a very common parasite of *Meloidogyne* and is often observed attached to nematode juveniles. The spore form can resist drought, exposure to fumigant nematicides and extreme temperature (Chen and Dickson, 1998). The inability to produce mass cultures of *P. penetrans* in quantities sufficient for large scale use is the major factor limiting practical work with these bacteria (Gowen and Ahmed, 1990). Dried tomato roots were then milled into a powder containing *Pasteuria* spores. The populations of *P. penetrans* which parasitize *Meloidogyne* not only prevent nematode reproduction, but also reduce infectivity of spore-encumbered juveniles. A reduction in infectivity may be evident when nematodes have as few as 15 spores attached (Davies *et al.*, 2000). Juveniles are prevented from invading roots when large enough numbers of spores are present in soil (Stirling, 1991).

Mechanisms of Infection

Pasteuria penetrans infects the root-knot nematode *Meloidogyne* spp. Spores of *Pasteuria* can attach to the cuticles of the second-stage juveniles, and germinate after the juvenile has entered roots and begun feeding. The germ tubes can penetrate the cuticle, and vegetative microcolonies then form and proliferate through the body of the developing female. Finally, the reproductive system of the female nematode degenerates and mature endospores are released into the soil.

Host Range

Attachment of the spores to the nematode cuticle is the first step in the infection process. However, spores of individual *Pasteuria* populations do not adhere to or recognize all species of nematode. The spores of each *Pasteuria* species usually have a narrow host range. For example, *P. penetrans* infects *Meloidogyne* spp., *P. thornei* infects *Pratylenchus* spp., and *P. nishizawae* infects the genera *Heterodera* and *Globodera*.

The processes associated with the initial binding of the endospores of *Pasteuria* spp. to their respective hosts have been explored by several laboratories. N-acetyglucosamine, which is present on the spore surface, is thought to be involved in adhesion by interacting with a lectin like receptor on the nematode cuticle. The nature of the cuticle receptor(s) for *Pasteuria* adhesion is ambiguous. Collagen may be responsible for the recognition process, because cuticle components involved in attachment are sensitive to trypsin and endoglycosidase F, and because gelatin (denatured collagen) itself can inhibit spore attachment (Preston *et al*., 2003).

Opportunistic Parasitic Bacteria

Bacteria live a saprophytic life, targetting nematodes as one possible nutrient resource. They are, however, also able to penetrate the cuticle barrier to infect and kill a nematode host in some conditions. They are described as opportunistic parasitic bacteria here, represented by *Brevibacillus laterosporus* strain G4 and *Bacillus* sp. B16.

Mechanism of Action

Brevibacillus laterosporus parasitizes the nematodes *Panagrellus redivius* and *Bursaphelenchus xylophilus*. After attaching to the epidermis of the host body, *B. laterosporus* can propagate rapidly and form a single clone in the epidermis of the nematode cuticle. The growth of a clone can result in a circular hole shaped by the continuous degradation and digestion of host cuticle and tissue. Finally, bacteria enter the body of the host, and digest all the host tissue as nutrients for pathogenic growth.

Host specificity of *Pasteuria penetrans* to nematodes

Bacteria	Nematode	Effect of bacteria on nematodes	References
P.penetrans	*M. incognita, Pratylenchus Brachurus*	Spore population of *M. incognita* and *P. brachyurus* attached only to *Meloidogyne* spp. and *P. brachyurus* respectively and not to the range of other nematodes tested	Dutky and Sayre, (1978)
P.penetrans	13 Nematode species	Large spores (4.3-6.6/~m) were associated with the endoparasitic nematodes	Spaull, (1981)
P.penetrans	*Heterodera elastic*	This isolate had distinct specificity within cyst nematodes	Nishizawa, (1984)
P. penetrans	*Meloidogyne* spp.	Spores of *P. penetrans* derived from *M. javanica* infected a much larger proportion of larvae of this species than the larvae of *M. incognita*. In contrast spores from *M. incognita* were equally effective	Spaull, (1984)
P.penetrans	*Meloidogyne* spp.	Specificity of *P. penetrans* isolate may be a response to nematode population rather than to nematode species	Stirling, (1991)
P.penetrans	*M. incognita, M. javanica*	J_2 were readily infested with spores of bacteria but varietal differences in susceptibility were observe	Siddiqui and Mahmood (1999)
P.penetrans	*Meloidogyne* spp.	Spores of *P. penetrans* from six populations of *Meloidogyne* only adhered to *Meloidogyne* and they adhered in greatest numbers to the species from which they were isolated	Davies *et al.* (1988)
P.penetrans	*Meloidogyne* spp.	The isolates varied greatly in their attachment to different nematode species and genera	Oostendorp *et al.* (1990)
P.penetrans	*Helicotylenchus pseudorobustus, Rotylenchus capensi*	Endospores of *P. penetrans* released from infected nematodes or adhering to their cuticle did not reveal significant morphological differences apart from endospores diameters	Volvas *et al.* (1993)
P.penetrans	*M. javanica, M. arenaria*	*P. penetrans* spores attached to *M. javanica* more readily than *M. arenaria* irrespective of the species of the host on which they had been produced and had no adaptation to the poorer hosts like *M. arenaria*	Channer and Gowen, (1994)
P.penetrans	*H. cajani*	The spores of this isolate had a mean diameter of 2.36/~m and were characterized using host range spore morphometric and serology range and spore morphometrics are not adequate for characterization of a species	Sharma and Davies, (1996)

During bacterial infection, the degradation of all the nematode cuticle components around the holes suggests the involvement of hydrolytic enzymes (Huang *et al.,* 2005).

Rhizobacteria

Another strategy used for the biological control of nematodes is based on the introduction of bacteria colonizing the rhizosphere of the host plant or so called rhizobacteria. These microorganisms that can grow in the rhizosphere also provide front line defence for roots against pathogen attack and are considered ideal for use as biocontrol agent (Weller, 1988). Rhizosphere bacteria have the ability to colonize plant roots and they also have positive effects on plant growth. They have been named plant growth promoting rhizobacteria (PGPR) by Kloepper *et al.* (1991) or plant health promoting rhizobacteria (PHPR) by Sikora (1993). Rhizobacteria have also been studied for the biological control of plant-parasitic nematodes. Aerobic endospore-forming bacteria (AEFB) (mainly *Bacillus* spp.) and *Pseudomonas* spp. are among the dominant populations in the rhizosphere that are able to antagonize nematodes. Other rhizobacteria reported to show antagonistic effects against nematodes include members of the genera *Actinomycetes, Agrobacterium, Arthrobacter, Alcaligenes, Aureobacterium, Azotobacter, Beijerinckia, Burkholderia, Chromobacterium, Clavibacter, Clostridium, Comamonas, Corynebacterium, Curtobacterium, Desulforibtio, Enterobacter, Flavobacterium, Gluconobacter, Hydrogenophaga, Klebsiella, Methylobacterium, Phyllobacterium, Phingobacterium, Rhizobium, Serratia, Stenotrotrophomonas* and *Variovorax*. Application of these rhizobacteria to sugar beet seed and potato seed pieces caused significant decreases in early root infection of the sugar beet cyst nematode *Heterodera schachtii* and potato cyst nematode *Globodera pallida* (Oostendorp and Sikora, 1990). Several rhizosphere bacteria with antagonistic activity against plant parasitic nematodes have been identified.

Zavaleta - Meija and Van Gundy, (1982) found that rhizobacteria have biocontrol activity towards root-knot nematode and they showed that more than 12% of the rhizobacteria tested reduced the number of galls of *M. incognita* on cucumber and tomato. Sikora, (1992) reported that 7-10% of the rhizosphere bacteria isolated from potato, sugar beet or tomato root systems have antagonistic activity against cyst and root knot nematodes. Sikora, (1993) revealed that PHPR influence the intimate relationship between the plant parasitic nematode and its host by regulation of nematode behaviour during the early root penetration phase of parasitism which is extremely important for crop yield.

Sikora, (1992) found that *Bacillus subtilis* was effective in controlling *M. incognita* on cotton and sugar beet, *M. arenaria* on peanut and *Rotylenchulus reniformis* on cotton. Strains of *Pseudomonas chitinolytica* also were shown to reduce *M. javanica* on tomato as reported by Spiegel *et al.* (1991). Smith, (1994) reported that *Bacillus* sp. strain 23a reduced *M. javanica* densities on tomato and *Pseudomonas fluorescens* strain Pf1 reduced the number of galls and egg masses of *M. incognita* on tomato roots (Santhi and Sivakumar, 1995). *B. cereus* strain S18 also decreased *M. incognita* on tomato as reported by Keuken, (1996).

Mechanism

The rhizobacteria usually comprise a complex assemblage of species with many different modes of action in the soil. Rhizobacteria reduce nematode populations mainly by regulating nematode behaviour, interfering with plant–nematode recognition, competing for essential nutrients, promoting plant growth, inducing systemic resistance or directly antagonizing by means of the production of toxins, enzymes and other metabolic products. The effects of these toxins include the suppression of nematode reproduction, egg hatching and juvenile survival, as well as direct killing of nematodes. Ammonia produced by ammonifying bacteria during decomposition of nitrogenous organic materials can result in reduced nematode populations in soil. *Corynebacterium paurometabolu* inhibited nematode egg hatching by producing hydrogen sulphide and chitinase.

Plant growth-promoting rhizobacteria (PGPR) can bring about ISR by fortifying the physical and mechanical strength of the cell wall by means of cell-wall thickening, deposition of newly formed callose, and accumulation of phenolic compounds (Preston *et al.,* 2003). They also change the physiological and biochemical ability of the host to promote the synthesis of defence chemicals against the challenge pathogen (e.g. by the accumulation of pathogenesis-related proteins, increased chitinase and peroxidase activity, and synthesis of phytoalexin and other secondary metabolites).

Bio-control efficacy of *P. fluorescens* through various modes of action

Studies with fluorescent pseudomonads demonstrated the control of soil borne pathogens through various modes of action.

Influence on Nematode Behaviour

Pathogen suppression depends mainly on competition for nutrients and space. Lectin binding structures of *P. fluorescens* associate with the root surface carbohydrates and interfere with the nematode host recognition and thereby suppressing root invasion by *H. schachtii* in sugarbeet. The production of toxic metabolites and the nematicidal components is reported to reduce hatching and invasion of the nematodes. The nematode behavior is controlled by the alteration of specific root exudates by *P. fluorescens.*

The nematode penetration is reduced due to lectin binding structure in the lipopolysaccharide layer of cell membrane. Application of *P. fluorescens* as seed dressing either in liquid or solid state resulted in 42 percent reduction in root penetration of *H. schachtii* on sugarbeet (Persidis *et al.*, 1991).

Siderophore Activity

The production of extracellular proteins termed as siderophores (fluorescent yellow green pigment) efficiently complexes or chelates the environmental iron, thereby limiting the availability of iron necessary for the growth and development of the pathogens and other soil microflora. Fluorescent pseudomonads have the property to form ferric siderophore complex, which prevents the availability of iron to the microorganisms. Ultimately this situation leads to iron starvation and prevents the survival of the microorganisms including nematodes. *P. aeruginosa* strain IE-6 and its streptomycin resistant strain IE-6^+ markedly suppressed nematode population densities in root and subsequent root-knot development and the iron concentration in soil was lowered by the addition of an iron chelator (Van Loon *et al.*, 1998; Ramamoorthy *et al.*, 2001).

Induction of Systemic Resistance (ISR)

ISR by definition refers to protection of the plants systemically by enhancement of plant's defensive capacity against a broad spectrum of pathogens or pests, which is acquired after appropriate inducing agent. *P. fluorescens* produces salicylic acid which is attributed for the induction of systemic resistance in host plants. Salicylic acid is an important systemic signal module in host plant. Plant growth promoting rhizobacteria equip the plant itself to combat nematodes. They impart resistance by sending some chemical signaling molecules so that the plants own defense mechanism is activated. Upon activation, the defense machinery responds by synthesizing certain enzymes and phenolics, which reinforce the cell wall structure which are directly inhibitory to the nematodes. This is often referred to as rhizobacteria mediated ISR (Van Loon *et al.*, 1998; Ramamoorthy *et al.*, 2001).

Salicylic acid (SA) dissolved at 10 mM in potassium –phosphate buffer (pH 7) and applied as foliar spray on tomato, inoculated with *M. incognita*, reduced nematode infestation and promoted plant growth. SA did not kill nematodes in *vitro* test. It enhanced synthesis of pathogenesis related –1 proteins as compared to uninoculated untreated and inoculated untreated plants. SA treatment has led to a significant increase in resistance to root knot nematode. Activities of the defense enzymes peroxidase, phenylalanine ammonia lyase, chitinase and catalase were significantly higher in bacterized tomato root tissues challenged with the nematodes. There was also significant reduction in the root- knot nematode population in roots and soil around the bacterized plants (Siddiqui and Shaukat, 2004).

Antagonistic Effect

P. fluorescens produces some nematicidal compounds, which affect the viability of the second stage juveniles of *M. incognita* and resulted in reduced galling in tomato and cucumber plants in green house. The antagonistic activity of *P. fluorescens* may be due to the alteration of bacterial root exudates which influences the nematode hatching, attraction, penetration and due to the production of secondary metabolites like antibiotics and cyanide. Antibiotics like pyrolnitrin, pyroverdine, pyoluteorine and 2,4-diaetyl pheloglucinol, hydrogen cyanide which degrade the toxin produced by the pathogen, salicylic acid and phytohormones like auxins, gibberellins and cytokinins stimulated plant growth and suppressed the deleterious microorganisms. Both cell concentrations and cell-free filtrates of Pf1 isolate suppressed egg hatching viability of immature females /males of *R. reniformis*. Significant mortality of the nematode was achieved at concentrations of 10^8 cells/ml (Van Loon *et al.*, 1998; Ramamoorthy *et al.*, 2001).

Effect on Enzymatic Activity

P. chitinolytica culture filterate exhibited strong chitinolytic, collagenalytic and proteolytic activities possessing lethal effect or causing juvenile mortality in *M. javanica.* Treatment of tomato seedlings with *P. chitinolytica* isolated form crustacean shell amended soil significantly decreased the penetration of J_2 of knot nematode into the roots and increased the plant growth. Lytic enzymes such as chitinases and β-1, 3- glucanases which degrade chitin (a major component of exoskeleton and peritrophic membrane of insect and nematodes) and glucan are produced which are lethal to the pathogens.

Chitinase enzyme produced by *P. fluorescens* hydrolyses the chitin layer on the eggshell during embryonic developement and thereby destroyed the developing embryo. Increase in the chitinase activity resulted in the suppression of phytonematodes in tobacco on application of *P. fluorescens*. Prior application of fluorescent pseudomonads protects the crop from the pathogens by strengthening the cell-wall structure and causing biochemical and physiological changes in the plant system (Sikora, 1993).

Nutrient Regulation

Phosphate solubilization, biological nitrogen fixation, improvement of other plant nutrients uptake and phytoharmone production like IAA are some of the mechanisms of PGPRs that directly influence plant growth. Strains *of P. fluorescens viz*., IISR-6, IISR-8, IISR-11, IISR-13 and IISR-52 obtained from the black pepper rhizosphere released inorganic phosphate from tricalcium phosphate. Significant uptake of nitrogen and potassium by the bacterized plants was noticed Siddiqui and Mahmood, (1999).

Plant Growth Promotion

The mechanisms involved in PGPR mediated plant growth promotion is directly by the production of plant growth regulators or indirectly by stimulating nutrient uptake by producing siderophores or antibiotics to protect plants from soil borne pathogens. These rhizobacteria produce gibberellin and mineralize phosphates.

Bacteria	Nematode	Effect	References
P.fluorescens	*M.javanica*	Reduced multiplication & morphometrics of females on nematode in different soils	Siddiqui and Mahmood (1999)
B. Subtilis & *B.pumilus*	*Meloidogyne* spp. *R. reniformis*	Reduced reproduction & galling on cotton, tomato & Peanut	Sikora, (1988)
B. thuringiensis	*Meloidogyne* spp.	Prevented gall formation on tomato.	Ignoffo and Dropkin (1977)
Streptomyces sp. *Isolate CR-43*	*Caenorhabditis elegans, M. incognita, Pratylenchus*	Reduced multiplication	Dicklow *et al.* (1993)
Clostridium butyricum	*Tylencho rhynchus martini*	Reduced nematode population	Johnston, (1958)
Agrobacterium Radiobacter	*Globodera pallida*	Resulted in reduced nematode infection by 40% when sprayed on seed pieces of potato	Sikora *et al.* (1988)

Cry protein-forming Bacteria

Bacillus thuringiensis (Bt) produces one or more parasporal crystal inclusions (Cry or d-endotoxins), which are known to be toxic to a wide range of insect species and some Cry proteins are also toxic to nematodes. There are six Cry proteins (Cry5, Cry6, Cry12, Cry13, Cry14, Cry21) known to be toxic to larvae of a number of free-living or parasitic nematodes (Kotze *et al.*, 2005).

Mechanism

Cry protein exerts its effects by forming lytic pores in the cell membrane of gut epithelial cells. After ingestion of toxin by target nematode larvae, the crystals dissolve within the gut of the nematode, and this is followed by proteolytic activation (Crickmore, 2005). Cry toxicity is directed against the intestinal epithelial cells of the midgut and leads to vacuole and pore formation, pitting, and eventual degradation of the intestine

(Marroquin *et al.*, 2000). The binding of poreforming toxin to a receptor in the epithelial cell is a major event. The bre gene encodes a glycosyltransferase, which is responsible for synthesizing a carbohydrate receptor glycolipid.

Endophytic Bacteria

Endophytic bacteria have been found internally in root tissue, where they persist in most plant species. They have been found in fruits and vegetables, and are present in both stems and roots, but do no harm to the plant. They have been shown to promote plant growth and to inhibit disease development and nematode pests. They found antagonistic properties towards *M. incognita* in 21 out of 181 endophytic bacteria. Several bacterial species have also been found to possess activity against root-lesion nematode (*Pratylenchus penetrans*) in soil around the root zone of potatoes. Among them, *M. esteraomaticum and K. varians* have been shown to play a role in root-lesion nematode suppression through the attenuation of host proliferation, without incurring any yield reduction. Rhizobacteria and endophytic bacteria display some of the same mechanisms for promoting plant growth and controlling phytopathogens, such as competition for an ecological niche or a substrate, production of inhibitory chemicals, and induction of systemic resistance (ISR) in host plants by Munif *et al.* (2000).

Endophytic bacteria have recently been found internally in the root tissue and in the rhizosphere where they persist in most plant species and plant tissues. They have been found in fruits and vegetables and in both stems and roots without doing harm to the plant (Hallmann, 2001). Recent research has demonstrated that bacterial endophytes can improve plant growth and reduce disease symptoms caused by several plant pathogens such as *Fusarium oxysporum* f. sp. *vasinfectum* on cotton (Chen *et al.,* 1996), *Verticillium albo-atrum* and *Rhizoctonia solani* on potato and on cotton (Kloepper *et al.,* 1992; Pleban *et al.,* 1995). Hallmann *et al.* (2001) found some evidence that endophytic bacteria may contribute to control of plant parasitic nematodes. They evaluated 7 isolates of endophytic bacteria isolated from cucumber and cotton roots against the root-knot nematode, *M. incognita* and they found a significant reduction of 50% in the number of galls on cucumber.

Munif *et al.* (2000) screened the endophytic bacteria isolated from tomato roots towards *M. incognita* on tomato under greenhouse conditions. They showed antagonistic properties in the screening of 21 out of 181 endophytic bacteria towards *M. incognita.*

Symbionts of Entomopathogenic Nematodes

Xenorhabdus spp. and *Photorhabdus* spp. are bacterial symbionts of the entomopathogenic nematodes *Steinernema* spp. and *Heterorhabdus* spp., respectively. They have been thought to contribute to the symbiotic association by killing the insect and providing a suitable nutrient environment for nematode reproduction. In recent years, a potentially antagonistic effect of the symbiotic complex on plant-parasitic nematodes has been reported (Lewis *et al.,* 2001). Further investigation demonstrated that the symbiotic bacteria seemed to be responsible for the plant-parasitic nematode suppression via the production of defensive compounds (Samaliev *et al.,* 2000). Three types of secondary metabolites have been identified as the nematicidal agent: ammonia, indole and stilbene derivative. They were toxic to second-stage juveniles of root-knot nematode (*M. incognita*) and to fourth-stage juveniles and adults of pine-wood nematode (*Bu. xylophilus*), and inhibited egg hatching of *M. incognita.*

References

Baoyu Tian, Jinkui Yang and Ke-Qin Zhang. 2007. Bacteria used in the biological control of plant-parasitic nematodes:populations,mechanisms ofaction, and future prospects. *FEMS Microbiol Ecol.*, 61 197–213.

Birchfield, W. and Antonpoulos, A. A. 1976. Scanning electron microscope observations of *Duboscquia penetrans* parasitizing root-knot larvae. *Journal of Nematology*, 8: 272-283.

Channer, A. G. and Gowen, S. R. 1992. Selection for increased host resistance and increased pathogen specificity in the *Meloidogyne-Pasteuria penetrans* interaction. *Fundamental & App. Nem.,* 15: 331-339.

Chen, Z. X. and Dickson, D.W. 1998. Review of *Pasteuria penetrans*: Biology, ecology, and biological control potential. *Journal of Nematology*, 30: 313-340.

Chen, Z. X., Dickson, D. W. and Hewlett, T. E. 1996. Quantification of endospore concentrations of *Pasteuria penetrans* in tomato root material. *J of Nem.,* 28: 50-55.

Crickmore, N. 2005. Using worms to better understand how *Bacillus thuringiensis* kills insects. *Trends Microbiol.,* 13: 347–350.

Davies, K. G. and Danks, C. 1993. Interspecific differences in the nematode surface coat between *Meloidogyne incognita* and *M. arenaria* related to the adhesion of the bacterium *Pasteuria penetrans. Parasitology*, 105: 475-480.

Davies, K.G., Fargette, M. and Balla, 2000. Cuticle heterogeneity as exhibited by *Pasteuria* spore attachment is not linked to the phylogeny of parthenogenetic root-knot nematode (*Meloidogyne* spp.). *Parasitol.,* 122: 111–120.

Dutky, E.M. and Sayre, R.M., 1978. Some factors affecting infection of nematodes by bacterial spore parasite *Bacillus penetrans. J of Nem.*, 10: 285.

Gowen, S. R. and Tzortzakakis, E. 1994. Biological control of *Meloidogyne* spp. with *Pasteuria penetrans. Bulletin OEPP*, 24: 495-500.

Gowen, S. R., and Ahmed, R. 1990. *Pasteuria penetrans* for control of pathogenic nematodes. *Aspects of Applied Biology*, 24: 25-32.

Hallmann, J. 2001. Plant interactions with endophytic bacteria. In press in M. J. Jeger, and N. J. Spence, Eds. Biotic interactions in plant-pathogen associations. CAB International, Wallingford, United Kingdom.

Huang, X.W., Tian, B.Y., Niu, Q.H., Yang, J.K., Zhang, L.M. and Zhang, K.Q. 2005. An extracellular protease from *Brevibacillus laterosporus* G4 without parasporal crystal can serve as a pathogenic factor in infection of nematodes. *Res Microbiol.,* 156: 719–727.

Kerry, B. R. 2000. Rhizosphere interactions and exploitation of microbial agents for the biological control of plant-parasitic nematodes. *Annu Rev Phytopathol.* 38: 423–441.

Keuken, O. 1996. Interaktionen zwischen dem Rhizosphärebakterium *Bacillus cereus*, dem Wurzelgallennematoden *Meloidogyne incognita* und tomate. Ph.D. disseratation, Bonn University, Bonn, Germany.

Kloepper, J.W., Rodriguez – Kabana, R., Mclnroy, J.A. and Collins, D.J. 1991. Analysis of population and physiological characterization of microorganisms in rhizosphere of plant with antagonistic properties to phytopathogenic nematodes. *Plant Soil,* 136: 95–102.

Kotze, A.C., O'Grady, J., Gough, J.M., Pearson, R., Bagnall, N.H., Kemp, D.H. and Akhurst, R.J. 2005. Toxicity of *Bacillus thuringiensis* to parasitic and free-living life stages of nematodes parasites of livestock. *Int J Parasitol.,* 35: 1013–1022.

Lewis, E.E., Grewal, P.S. and Sardanelli, S. 2001. Interactions between *Steinernema feltiae–Xenorhabdus bovienii* insect pathogen complex and root-knot nematode *Meloidogyne incognita*. *Biol Contr.,* 21: 55–62.

Marroquin, L.D., Elyassnia, D., Griffitts, J.S., Feitelson, J.S. and Aroian, R.V. 2000. *Bacillus thuringiensis* (Bt) toxin susceptibility and isolation of resistance mutants in the nematode *Caenorhabditis elegans*. *Genet.,* 155: 1693–1699.

Meyer, S. L. F. 2003. United States Department of Agriculture – Agricultural Research Service research programs on microbes for management of plant-parasitic nematodes. *Pest Manag Sci.* 59: 665–670.

Munif, A., Hallmann, J. and Sikora, R. A. 2000. Evaluation of the biocontrol activity of endophytic bacteria from tomato against *Meloidogyne incognita*. Mededelingen Faculteit Landbouwkundige, Universiteit, *Gent.,* 65: 471-480.

Nishizawa, T., 1984. Effects of two isolates of *Bacillus penetrans* for control of root-knot nematodes and cysts nematodes. *Proceedings, First International Congress of Nematology Guelph*, Canada, p.60.

Oostendorp, M. and Sikora, R. A. 1990. *In vitro* interrelationships between rhizosphere bacteria and *Heterodera schachtii*. *Revue de Nematologie.,* 13: 269-274.

Persidis, A., Lay, J.G., Manousis, T., Bishop, A.H. and Ellar, D.J. 1991. Characterization of potential adhesions of the bacterium *Pasteuria penetrans*, and of putative receptors on the cuticle of *Meloidogyne incognita*, a nematode host. *J Cell Sci.,* 100: 613–622.

Preston, J.F., Dickson, D.W., Maruniak, J.E., Nong, G., Brito, J.A., Schmidt, L.M. and Giblin-Davis, R.M. 2003. *Pasteuria spp.:* systematics and phylogeny of these bacterial parasites of phytopathogenic nematodes. *J Nematol.,* 35: 198–207.

Ramamoorthy, V., Viswanathan, R., Raguchander, T., Prakasam, V. and Samiyappan, R. 2001. Induction by systemic resistance by plant growth promoting rhizobacteria in crop plants against pests and diseases. *Crop Protec.,* 20: 1–11.

Samaliev, H.Y., Andreoglou, F.I., Elawad, S.A., Hague, N.G.M. and Gowen, S.R. 2000. The nematicidal effects of the bacteria *Pseudomonas oryzihabitans* and *Xenorhabdus nematophilus* on the root-knot nematode *Meloidogyne javanica*. *Nematol.,* 2: 507–514.

Santhi, A. and Sivakumar, V. 1995. Biocontrol potential of *Pseudomonas fluorescens* (Migula) against root-knot nematode, *Meloidogyne incognita* (Kofoid and White) Chitwood, 1949 on tomato. *J of Biol Cont.,* 9:113-115.

Sayre, R. M. and Starr, M. P. 1985. *Pasteuria penetrans*, a mycelial and endospore-forming bacterium parasitic in plant-parasitic nematodes. *Proceedings of the Helminthological Society of Washington,* 52:149-165.

Sharma, S. B. and Davies, K. G. 1996. Characterization of *Pasteuria* isolated from *Heterodera cajani* using morphology, pathology, and serology of endospores. *Systemic and App. Microbiol.,* 19:106-112.

Siddiqui, I.A. and Shaukat, S.S. 2004. Systemic resistance in tomato induced by biocontrol bacteria against the root-knot nematode, *Meloidogyne javanica* is independent of salicylic acid production. *J Phytopathol.,* 152: 48–54.

Siddiqui, Z.A. Mahmood, I. 1999. Role of bacteria in the management of plant parasitic nematodes"A review. *Bioresource Tech.*, 69: 167-179.

Sikora, R. A. 1988. Interrelationship between plant health-promoting rhizobacteria, plant parasitic nematodes and soil microorganisms. *Mededelingen Faculteit Landbouwkundige Rijksuniversiteit Gent.,* 53: 867-878.

Sikora, R. A., Reckhaus, P. and Adamou, I. 1988. Presence, distribution and importance of plant parasitic nematodes in irrigated agricultural crops in Niger. *Mededelingen van de Faculteit Landbouwwetenschappen, Rijksuniversiteit Gent,* Belgium, 53: 821-834.

Sikora, R.A. and Hoffmann-Hergarten, S. 1993. Biological control of plant parasitic nematodes with plant-health promoting rhizobacteria. Biologically based technology (Lumsden PD & VaughJL, eds), pp. 166–172.

Sikora, R.A. 1992. Management of the antagonistic potential in agriculture ecosystems for the biological control of plant parasitic nematodes. *Annu Rev Phytopathol.,* 30: 245–270.

Smith, L. J. 1994. The potential of rhizobacteria as biological control agents of root-knot nematode. Thesis, University of Queenland, Australia.

Spaull, V.W., 1981. *Bacillus penetrans* in South African plant parasitic nematodes. *Nematologica,* 27: 244-245.

Spaull, V.W., 1984. Observations of *Bacillus penetrans* infecting *Meloidogyne* in sugarcane fields in South Africa. *Revue Nem.*, 7: 277-282.

Spiegel, Y., Cohn, E., Galper, S., Sharon, E. and Chet, I.1991. Evaluation of a newly isolated bacterium, *Pseudomonas chitinolytica* for controlling the root-knot nematode *Meloidogyne javanica*. *Biocont Sci and Tech.*, 1:115-125.

Starr, M. P., and Sayre, R. M. 1988. *Pasteuria thornei* sp. nov. and *Pasteuria penetrans* sensu stricto emend., mycelial and endospore-forming bacterium parasitic, respectively, on plant-parasitic nematodes of the genera *Pratylenchus* and *Meloidogyne*. *Annals de l'Institut Pasteur, Microbiologie,* 139: 11-31.

Stirling, G.R. 1991. Biological Control of Plant Parasitic Nematode: Progress, Problems and Prospects. *CAB International*, Wallington, UK.

Van Loon, L.C., Bakker, P.A.H.M. and Pieterse, C.M. 1998. Systemic resistance induced by rhizosphere bacteria. *Annu Rev Phytopathol* 36:453–483.

Weller, D. M. 1988. Biological control of soil-borne plant pathogens in the rhizosphere with bacteria. *Ann Review of Phytopathol.*, 26: 379-407.

Zavaleta-Meija, E., and Van Gundy, S. D. 1982. Effects of rhizobacteria on *Meloidogyne* infection. *J of Nem.,* 14: 475A-475B.

3

Fungal Antagonists Against Phytonematodes

A.S. Krishnamoorthy and R. Priyanka

Department of Plant Pathology, Tamil Nadu Agricultural University Coimbatore- 641 003, Tamil Nadu

Introduction

Plant parasitic nematodes are important pests, causing major damage to the world's food and fibre crops. The management of nematodes is more difficult than that of other pests because nematodes mostly inhabit the soil and usually attack the underground parts of the plants. Nematodes in soil are subject to infections by bacteria and fungi. This creates the possibility of using soil microorganisms to control plant-parasitic nematodes. Biocontrol of nematodes was first studied by Duddington (1951). The development of biological control agents is also considered an effective alternative for nematode control on vegetables (Van Gundy, 1985; Kerry, 1987). There are more than 200 species of nematophagous fungi described. Based on the infection mechanism, they are commonly subdivided into three main groups: the nematode-trapping fungi that captures free living nematodes using specialized morphological structures (i.e., traps), the endoparasitic fungi that infects nematodes using adhesive spores, and the egg- and cyst-parasitic fungi that infect these stages with their hyphal tips (Barron, 1977). The nematophagous fungi that have received most attention for biological control of plant-parasitic nematodes include various species of the nematode-trapping fungi *Arthrobotrys* spp. (Stirling and Smith, 1998), and the egg-parasitic fungi *Pochonia chlamydosporium* (Kerry, 2001) and *Paecilomyces lilacinus* (Gaspard *et al*., 1990).

Nematode Trapping Fungi

Nematode trapping fungi are also known as predatory fungi have different types of modified organ called traps that are capable of trapping the nematodes.

Trapping organs of predatory nematophagous fungi are adhesive nets simple, complex, adhesive spore sessile and stalked, simple two-dimensional adhesive networks, adhesive knobs and branches, non-constricting rings, constricting rings open and closed (Gray, 1988).

a. Sticky branches

- Fungal hyphae – short lateral branches (1 – 3 celled) covered with adhesive substance
- Anastamose to form loops
- Nematodes are trapped by these loops.

  Eg. *Monacrosporium cionopagum, M. gephyropagum, Dactylella lobata*

b. Sticky networks

- Mycelium curls around and anastamosis with similar branches.
- Loops produces complex three dimensional and two dimensional structures of diameter 20-59 μm.
- Adhesive surface of networks to hold the nematode.

  Eg. *A. superba, A. oligospora, Arthrobotrys species.*

c. Sticky knobs

- Small spherical or sub-spherical lobes are present on 1 or 2 celled lateral hyphae.
- Only the terminal knob is sticky to hold the nematodes.

  Eg. *Monacrosporium ellipsosporum, Dactylaria candida, Dactylella leptospora*

d. Constricting rings

- The short hyphal branch curls back on itself and anastamosis and forming a ring.
- When the nematode enters the ring and contact the inner walls of ring cells.
- Ring cells bulge inward, filling the lumen of the ring and kills the nematode.

  ‘Eg. *Arthrobotrys dactyloides, Dactylaria brochopaga, Monacrosporium doedycoides*

e. Non – Constricting rings

- Trap is formed similar to the constricting ring
- Non – adhesive trap
- Ring becomes an infective structure which kills the nematode

  Eg. *Dactylaria candida* Chemotrophic growth of nematophagous fungi hyphae towards nematodes was observed in detached traps of *Dactylella leptospora*

Female and Egg Parasitic Fungi

This group of fungi parasitizes nematode females and/or eggs. These fungi are facultative parasites such as, *Verticillium* spp., one of the most important pathogens of root-knot and cyst nematodes. Most species in this genus have been recorded as occurring in cysts and/or eggs of *Globodera*, *Heterodera* and *Meloidogyne* (Morgan-Jones and Rodriguez-Kabana, 1988).

V. chlamydosporium is the most widely studied species. These fungi form branched mycelia networks which when in close contact with the egg shell, penetrated the egg wall and destroy its contents (Lopez-Llorca and Duncan, 1988). Morgan-Jones *et al.*, (1983) and Meyer *et al.*, (1990) suggested that *V. chlamydosporium* itself might produce a toxin that affects egg hatch.

Paecilomyces has been one of the principal genera in biological control in recent years. Lysek (1976) was the first to observed it in association with nematode eggs. Jatala *et al.*, (1979) found it parasitizing eggs of *M. incognita* and *P. lilacinus* as a biological control agent against several nematode species. This fungus penetrates the nematode eggs directly by individual hyphae (Morgan-Jones *et al.*, 1984). The mycoparasitic ability of *Trichoderma sp.* against soil-borne plant pathogens allows for the development of biocontrol strategies. Windham *et al.* (1989) reported reduced egg production in the root-knot nematode *Meloidogyne arenaria* following soil treatment with *Trichoderma harzianum* and *T. koningii* preparations. Combining *T. harzianum* with neem cakes reduced the population of citrus nematode, *Tylenchulus semipenetrans* (Parvatha *et al.*, 1996). Reduction of *M. javanica* infection with several isolates of *Trichoderma lingnorum* and *T. harzianum* has been reported (Spiegel and Chet, 1998).

Dactylella oviparasitica was the first parasite found in *Meloidogyne* eggs (Stirling and Mankau, 1978). This fungus produces appressoria on the egg surface and then penetrates the eggs through these specialized structures and through enzymatic penetration especially by chitinase production (Stirling *et al.*, 1979).

The habit of nematode trapping fungi, especially the constricting ring forming fungus *D. brochopaga* that form traps immediately upon germination in the soil environment. When nematodes pass through the constricting rings of *D. brochopaga*, the ring cells are inflated inwards due to friction of the nematode body, and consequently they are captured and strangulated by these rings (Persmark and Nordbring-Hertz, 1997). The ring cells swell immediately after nematode entry and the captured nematode struggles for life till death. After the death of the nematode, the fungus grows inside and eventually consumes the whole content of the nematode body (Singh and Bandyopadhyay, 2001). The sophisticated capturing mechanism and higher predacity of saprophytic as well as parasitic nematodes indicates the biocontrol potential of the fungus. This fungus was effective against some nematodes particularly the second-stage juveniles (J2) of *Meloidogyne* sp. and *Heterodera* sp. (Singh *et al.*, 2006).

Pochonia chlamydosporia infects nematode eggs through the development of appressoria at the hyphal tip or laterally, which appear to attach tightly to the surface of eggshells, which are penetrated by an infection peg. A post-infection bulb leads to the development of a mycelium within the egg and the destruction of its contents (Segers *et al.*, 1996).

Mycorrhizal Fungi

The mycorrhizas are symbiotic associations between plant roots and certain species of fungi. In practical agricultural terms, mycorrhiza are traditionally ectomycorrhiza and vesicular arbuscular mycorrhiza as reported by Ikram (1990). A plant with a well-established symbiont density is stronger because it has:

a) Increased resistance to nematode parasities and root pathogens

b) Increased tolerance to drought, salt and other abiotic stress factors

c) Improved phosphate uptake

d) Less root injury following transplanting

The colonization of plants with endomycorrhizal fungi apart from providing plants with nutrients also has a depressive effect on root-knot nematodes. The obligate symbiotic endomycorrhizal fungi protect their host against root-knot nematode attack by competition as reported by Sikora (1978) and Hussey & Roncadori (1982).

Sikora (1978) found that penetration and development of *M. incognita* in tomato was significantly reduced by the endomycorrhizal fungus, *Glomus mosseae* under glasshouse conditions. Because these fungi occur commonly together with plant parasitic nematodes in the roots or rhizosphere of the same plants, they interact with both host plant and nematodes (Stirling, 1991).

Smith (1987) explained possible hypotheses for the beneficial effects of endomycorrhizal fungi on plants parasitized by nematodes. He postulated that the symbiotic mycorrhizae:

(1) Alter root exudates consequently affecting egg hatching or nematode attraction

(2) Retard nematode development or reproduction within root tissue

(3) Parasitize female nematodes and their eggs.

Endophytic Fungi

Another group of fungi that has been recently used as a biocontrol agent against plant parasitic nematodes especially root-knot nematodes, are endophytic fungi (Hallmann *et al.*, 1997). These saprophytic fungi colonize healthy plant tissue without causing symptoms. When the colonization is successful it leads to protection of the plant against biotic and/or abiotic stress, these fungi are called mutualistic endophytic fungi (Carroll, 1990). Mousa and Hague (1988) found that in the soybean disease complex between *Fusarium oxysporum f. sp. glycines* and *M. incognita* active colonization of the giant cells by the fungus resulted in reduced development of juveniles and an increased proportion of males.

Hallmann and Sikora (1993) evaluated 200 isolates of endophytic fungi, representing different genera, isolated from tomato roots. Forty isolates were screened for their ability to control *M.incognita* in pot experiments. Hallmann and Sikora (1994, 1995) found a reduction in gall formation by *M. incognita* between 52 and 75% after application of four endophytic strains of the fungus *Fusarium oxysporum*. They also found that *M. incognita* attraction and penetration of tomato seedlings was significantly reduced following treatment with the culture filtrate of *Fusarium oxysporum*.

References

Barron, G. L.1977. The nematode destroying fungi. Canadian Biological Publications Ltd, Guelph.

Carroll, G. C. 1990. Fungal endophytes in vascular plants: Mycological research opportunities in Japan. *Transactions Mycological Society Japan,* 31:103-116.

Duddington, C. L. 1951. Two new predacious hyphomycetes. *Transactions of the British Mycological Society,* 34:598-603.

Gaspard, J. T., Jaffe, B. A., Ferris, H. 1990. *Meloidogyne incognita* survival in soil infested with *Paecilomyces lilacinus* and *Verticillium chlamydosporium. J. Nematol.,* 22:176–181.

Gray, N. F. 1988. Fungi attacking vermiform nematodes. In: Poinar, G. O., and Jansson, H.B., Eds. Diseases of nematodes, Vol. II, CRC Press Inc., Boca Raton. Pp. 3-38

Hallmann, J., and Sikora, R. A. 1993. Fungal endophytes and their potential for biological control of root-knot nematodes. Proceedings of 6th International Congress of Plant Pathology, Montreal, Canada.

Hallmann, J., and Sikora, R. A. 1994. Influence of *Fusarium oxysporium*, a mutualistic fungal endophyte on *Meloidogyne incognita* infection of tomato. *Journal of Plant Diseases and Protection,* 101:475-481.

Hallmann, J., and Sikora, R. A. 1995. Occurrence of plant parasitic nematodes and nonpathogenic species of *Fusarium* in tomato plants in Kenya and their role as mutualistic synergists for biological control of root-knot nematodes. *International Journal of Pest Management,* 40:321-325.

Hallmann, J., Rodriguez-Kabana, R. and Kloepper, J. W. 1997. Nematode interactions with endophytic bacteria. In: A. Ogoshi, K. Kobayashi, Y. Homma, F. Kodama, N. Kondo, and S. Akino, Eds. Plant growth-promoting rhizobacteria-present status and future prospects. Nakanishi printing, Sapporo. Pp. 243-245

Hussey, R. S. and Roncadori, R. W. 1982. Vesicular-arbuscular mycorrhizae may limit nematode activity and improve plant growth. *Plant Disease,* 66:9-14.

Ikram, A. 1990. Beneficial soil microbes and crop productivity. The Planter. 66:640-648.

Jatala, P., R. Kaltenbach, and M. Bocangel. 1979. Biological control of *Meloidogyne incognita acrita* and *Globodera pallida* on potatoes. Journal of Nematology, 11:303.

Kerry, B. R. 1987. Biological control. In: Brown, R. H., and Kerry, B. R., Eds. Principles and practice of nematode control in crops. Academic Press, New York, USA. Pp. 233-263

Kerry, B. R. 2001. Exploitation of the nematophagous fungus *Verticillium chlamydosporium* Goddar for the biological control of root-knot nematodes (*Meloidogyne* spp.). In: Butt TM, Jackson C, Magan M (Eds) Fungi as biocontrol agents; progress, problems and potential. CABI Publishing, Wallingford, Pp. 155–168

Lopez-Llorca, L. V., and Duncan, G. H. 1988. A study of fungal endoparasitism of the cereal cyst nematode *Heterodera avenae* by scanning electron microscopy. Canadian *Journal of Microbiology,* 34:613-619.

Lysek, H. 1976. Autodehelminthization of soil in lowland deciduous forests. Universitatis Palackianae Olomucensis Facultatis Medicae, 41:73-106.

Meyer, S. L. F., Huettel, R. N. and Sayre, R. M. 1990. Isolation of fungi from *Heterodera glycines* and in vitro bioassays for their antagonism to eggs. *Journal of Nematology*, 22:532- 537.

Morgan-Jones, G. and Rodriguez-Kabana, R. 1988. Fungi colonizing cysts and eggs. In: Poinar, G. O., and Jansson, H. B., Eds. Diseases of nematodes, Vol. II, CRC Press Inc., Boca Raton. Pp. 39-58

Morgan-Jones, G., White, J. F. and Rodriguez-Kabana, R. 1983. Phytonematode pathology: ultrastructural studies. I. Parasitism of *Meloidogyne arenaria* eggs by *Verticillium chlamydosporium*. *Nematropica,* 13:245-260.

Morgan-Jones, G., White, J.F. and Rodriguez-Kabana, R. 1984. Phytonematode pathology:ultrastructural studies. I. Parasitism of *Meloidogyne arenaria* eggs and larvae by *Paecilomyces lilacinus*. *Nematropica,* 14:57-71.

Mousa, E. M. and Hague, N. G. M. 1988. Influence of *Fusarium oxysporium* f. sp. *glycines* on the invasion and development of *Meloidogyne incognita* on soybean. *Reveu de Nematologie,* 11:437-439.

Parvatha, R. P., Rao, M. S. and Nagesh, M. 1996. Management of citrus nematode, *Tylenchulus semipenetrans* by integration of *Trichoderma harzianum* with oil cakes. *Nematol. Medit.,* 24: 265-267.

Persmark, L. and Nordbring-Hertz, B. 1997. Conidial trap formation of nematode-trapping fungi in soil and soil extracts. FEMS *Microbiol. Ecol.*, 22:313–323.

Segers, R., Butt, T. M. and Kerry, B. R. 1996. The role of the proteinase VCP1 produced by the nematophagous *Verticillium chlamydosporium* in the infectionprocess of nematode eggs. *Myco. Res.,* 100:421–428.

Sikora, R. A. 1978. Einfluß der endotrophen Mykorrhiza *Glomus mosseae* auf das Wirt-Parasit-Verhältnis von *Meloidogyne incognita* an Tomaten. *Journal of Plant Disease and Protection,* 85:197-202.

Singh, K. P. and Bandyopadhyay, P. 2001. Induction of trapping rings and assessment of predacity of *Dactylaria brochopaga* Drechsler against some plant parasitic nematodes. *Indian. Phytopathol.*, 54:223–239.

Singh, K. P., Jaiswal, R. K., Kumar. N. and Kumar, D. 2006. Nematophagous fungi with root-galls of rice caused by *Meloidogyne graminicola* and its control by *Arthrobotrys dactyloides* and *Dactylaria brochopaga*. *J. Phytopathol.*, 154:1–5.

Smith, G. S. 1987. Interactions of nematodes with mycorrhizal fungi. In J. A. Veech, and D. W. Dickson, eds. Vistas on nematology. Hyatlsville, Maryland. Pp. 292-300.

Spiegel, Y. and Chet, I. 1998. Evaluation of *Trichoderma* spp. as a biocontrol agent against soil-borne fungi and plant parasitic nematodes in Israel. *Integrated Pest Management Reviews,* 3: 169–175.

Stirling, G.R. 1991. Antagonists of nematodes. G. R. Stirling, Eds. Biological control of plant parasitic nematodes. CAB International, Wallingford, Oxoid, UK. Pp. 50-98.

Stirling, G.R., Mankau, R. 1978. Parasitism of *Meloidogyne* eggs by a new fungal parasite. *J. Nematol.,* 10:236–240.

Stirling, G. R., McKenry, M. V. and Mankau, R. 1979. Biological control of root-knot nematodes (*Meloidogyne* spp.) on peach. *Phytopathology,* 69:806–809

Stirling, G.R., Smith, L.J. 1998. Field tests of formulated products containing either *Verticillium chlamydosporium* or *Arthrobotrys dactyloides* for biological control of root-knot nematodes. *Biol. Control.*, 11:231–239

Van Gundy, S. D. 1985. Biological control of nematodes: status and prospect in agricultural IPM systems. In Biological Control in Agricultural IPM Systems. Academic Press, London. Pp. 467-477

Windham, G. L., Windham, M. T. and Williams, W. P. 1989. Effects of *Trichoderma* spp. onmaize growth and *Meloidogyne arenaria* reproduction. *Plant Dis.,* 73: 493–494.

4

Mass Production of Fungal and Bacterial Antagonists

S. Nakkeeran, G. Karthikeyan and S.Vinod Kumar

Department of Plant Pathology, Tamil Nadu Agricultural University Coimbatore- 641 003, Tamil Nadu

Introduction

The need of the agricultural community was the force of motivation for the development of broad-spectrum pesticides towards the management of pests and diseases of cultivated crops. Advances in pesticide were measured exclusively by the assessment of its efficacy in management alone. But today food production has ceased to be a major concern, due to stable population levels and success of innovative agricultural technologies. At this juncture, the agrochemical industry that primarily recognized the farmers as their customers has failed to acknowledge the public as an important client. Perception of the public reflects that, the negative aspects of pesticides seem to outnumber their benefits. The increased reflection on environmental concern over pesticide use has been instrumental in a large upsurge of biological disease control. Development of fungicide resistance among the pathogens, ground water and foodstuff pollution and the development of oncogenic risks has further encouraged for the exploitation of antagonistic microflora in disease management

Biological control of plant pathogens can be broadly defined as the reduction of inoculum density or disease producing activities of a pathogen or parasite in its active or dormant state by one or more organisms, accomplished naturally or through manipulation of environment, host or antagonist or by mass introduction of one or more antagonists (Baker and Cook, 1974). The decades of laboratory experiments have led to the explosion in the field of bio-control and introduced more than 50 commercial formulations comprising of fungi and bacteria (Whipps, 1997). Among the antagonists *Trichoderma,* fluorescent *Pseudomonas,*

Bacillus spp., and rhizobacteria plays a vital role in the management of fungal nematode complex. Hence, in this chapter we discuss in detail about the mass production techniques for both fungal and bacterial antagonists.

Trichoderma as a Successful Bio-control Organism

Trichoderma is the most widely used biocontrol agent. Among the various isolates of *Trichoderma, T.viride, T.harzianum, T.virens* and *T.hamatum* are used against the management of various diseases of crop plants.

It has many advantages as bio-control agent

a) High rhizosphere competence

b) High competitive saprophytic ability

c) Enhance plant growth

d) Great arsenal of inducible polysaccharide –degrading enzymes

e) Ease for mass multiplication

f) Broad spectrum of action against various pathogens

g) Provide excellent and reliable control

h) Environmentally safe

Equipments Used for Mass Production

Bright field microscope

Laminar air Flow Chamber

Laminar air flow chamber is used for the culturing of microorganisms under aseptic conditions. It works by passing sterile air into the chamber using the fibrous filters. Before the usage of the equipment clean the work bench with surface disinfectant and then close the doors of the chamber and switch on the UV lamp for a period of 20minutes.Then switch off the UV lamp and switch on the air flow and the lights. Then open the chamber for use.

Autoclave

The autoclave is an apparatus, used for the sterilization of culture media (autoclaving). The increased pressure increases the boiling point of water and produces steam with high temperature which destroys the cells of microbes. Autoclave and pressure cooker works on the same principle (Boyles law). Potato dextrose agar/broth and molasses yeast medium have to be sterilized at 15 lb / sq.inch (121°C) for 20 minutes.

Microwave Oven

It is mainly used for boiling and melting of culture media very quickly. Boiling of potato, *Agar agar* can be done efficiently within a short time. Care should be taken that only glass / oven proof vessels alone must be used. Avoid metal vessels.

Hot Air Oven

It consists of an insulated cabinet held at a constant temperature by means of an electric heating mechanism and thermostat. It is fitted with a fan to keep the hot air circulating at a constant temperature and thermometer for recording the temperature of the oven. For normal sterilization work, the oven should be operated at 160°C. Glass wares are to be sterilized for 2 hours at 160°C.

Fermenter

Fermenter is equipment which is used for mass multiplication of microorganism. There are different models with varying capacities. The fermenter shown in the figure consists of two basic units. Unit 1 consists of steam generator and unit 2 consists of fermenter vessel. The stream generator is a closed container which consists of two heating elements of 2.0 KW of each totaling 4.0 KW capacity fitted at the bottom of the vessel. On the top of the vessel is the pressure gauge, thermostat, safety valve and delivery tube are attached.

The second part consists of triple layered stainless steel container, contains the fermenter vessel in the centre and a rotor is fitted to it which is operated by the use of motor fitted at the bottom. The fermenter vessel is closed air tightly with a lid fitted with inlet port, outlet port, pressure gauge, safety valve and steam release outlet.

Other equipments/ glassware required for culturing work

1. Petridish holder (Copper jacket)
2. Inoculation needle
3. Spirit lamp/ Bunsen burner
4. Haemocytometer
5. Balance
6. Test tube stand
7. Pipette
8. Petriplates

9. Test tubes
10. Conical flasks

Mass Production and Formulations

The first major concern in commercial production systems involves the achievement of adequate growth of the biocontrol agent. In many cases biomass production of the antagonist is difficult due to the specific requirement of nutritional and environmental conditions for the growth of organism. Mass production is achieved through liquid and semisolid and solid fermentation techniques. The commercial success of biocontrol agents requires

- Economical and viable demand
- Consistent and broad spectrum action
- Safety and stability
- Location specific strains
- Longer shelf life
- Low capital costs

Liquid Fermentation

This fermentation system has been adopted for the mass multiplication of fungal and bacterial biocontrol agents. For mass multiplication the selected medium should be inexpensive and readily available with appropriate nutrient balance. Acceptable materials include molasses, corn steep liquor and sulphate waste liquor, and jaggery. Small-scale fermentation in molasses yeast medium resulted in the abundant chlamydospore production. Preparations of *Trichoderma* containing chlamydospores were more effective in preventing the diseases rather than that of the conidia based formulations (Papavizas and Lewis, 1989).

Solid Fermentation

In nature wide range of organic substrates with low microbial are available and could be used for the solid-state fermentation for mass multiplication. Large number of composted organic products has been studied for their use as horticultural substrates. These often have to be carefully utilized to remove the toxic substances and may contain pathogenic propagule with high microbial load. Hence the media with relatively low microbial content would be suited for solid-state fermentation and beneficial biological amendment. Solid substrates for the production of the inoculums of various biocontrol fungi include straws, wheat bran, sawdust, moistened bagasse, sorghum grains, paddy chaff, decomposed coir pith, farmyard manure and other substrates rich in cellulose.

Semi-solid Fermentation

It is used for the fungus, which do not sporulate in liquid state fermentation. The chances of contamination are high when compared to liquid state fermentation.

Talc Based Formulation

The Tamil Nadu Agricultural University, Coimbatore, has developed the technology of mass production of talc-based formulation of *Trichoderma viride*, and *Pseudomonas fluorescens* for seed treatment. Annually it produces this formulation to cover 20,000 ha. Several private industries in Coimbatore, Banglore, Chethali, Delhi and Chennai produce the same in large quantities. However, the demand exceeds the present supply. The annual requirement of *Trichoderma* has been estimated as 5,000 tones to cover 50 per cent area in India. Indirectly it also creates self-employment opportunities to the unemployed youths. The first commercial seed treatment formulation of bio-control agent in India was developed by Jeyarajan *et al.* (1994).

T. viride was grown in molasses yeast medium for 10 days in flasks. Then 2.51 of the culture were inoculated to 501 of sterilized molasses yeast medium in fermentor. It was incubated for 10 days with 4-8 hr aeration / day. The fungal biomass and the broth were mixed with 100-kg talc (super white) powder and 500 g of carboxy methylcellulose as a sticker. It was dried in shade for 72 hr and packed in alkathene bags. The initial population of *Trichoderma* in the produce was 300 x 10^6 cfu/g. The product should contain a minimum of 20 x 10^6 cfu/g at the time of use. This is applied at the rate of 4 g per kg of seed. The shelf life was 4 months. Angappan (1992) found that in chickpea treated with this product the rhizosphere population was maintained at 11 to 13 x 10^3 cfu per gram throughout crop growth stage.

Development of Trichoderma Product

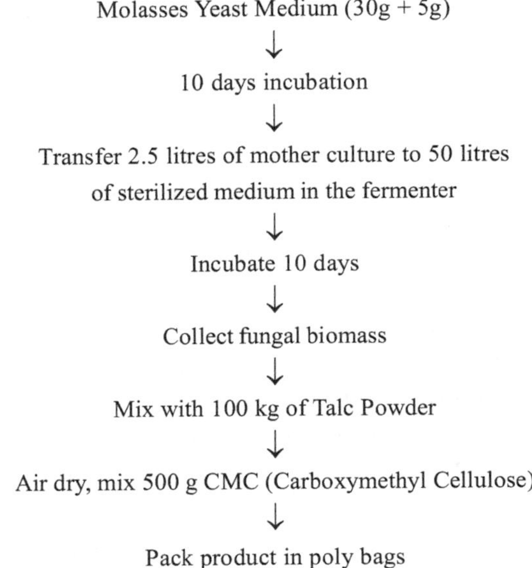

Gypsum

Ranganathan *et al.* (1995) found that gypsum was a good substitute for talc but was much cheaper than talc.

Industrial Wastes

Nakkeeran and Jeyarajan (1996) tested two industrial waste namely precipitated silica and calcium silicate in the place of talc and found them to give a population of 99, 104 x 10^6 cfu/g respectively compared to 143 x 10^6 cfu/g in talc substrate after 4 months of storage. These two substrates were also much cheaper than talc.

Diatomaceous Earth Granules

Diatomaceous earth granules were added into a broth consisting of 100-ml black grap molasses, 900 ml of water, 3 g each of KNO_3 and KH_2PO_4 till the level of saturation. It was autoclaved at 121°Cfor 15 minutes and spread in shallow pans to a height of 3-5 cm and autocalved again. *T. harzianum* (3 day old culture) was homogenized in a blender for 3 seconds and mixed with sterilized granules in shallow pans and incubated at 25°C for 4-7 days. Clumps were broken up and granules air dried with frequent stirring and used as an inoculum. The inoculum was mixed at 1.1 (v/v) with sterilized diatomaceous earth granules impregnated with 10% molasses solution. It was applied to peanut at 140 kg/ha on 70 and 100 days after sowing to control *Sclerotium rolfsii*. The disease was reduced by 42% over control and yield increased by 13.5% (Backman and Rodriguez Kabana, 1975).

Wheat Bran: Saw Dust Formulation

Wheat bran: Sawdust: tap water mixture (3:1:4 v/v) is taken in polypropylene bags and autoclaved for 1 hour at 121°C for two successive days. The bags were inoculated with *Trichoderma harzianum* and incubated in illuminated chambers for 14 days at 30°C. It was applied at the time of sowing and mixed with the soil to a depth of 7-10 cm with a rotary hoe. It increased yield of beans (1500 kg/ha), tomato (300 kg/ha), cotton (500 kg/ha) and potato (40-600 kg/ha) and controlled *Sclerotium rolfsii* and *Rhizoctonia solani* (Elad *et al.*, 1986).

Wheat bran: Peat

Wheat bran : Peat mixture (1:1 v/v) was autoclaved for 1 hour. Substrate moisture was adjusted to 50% (W/W) with sterile water, medium was inoculated with 0.1 ml of a conidial suspension containing 2 x 10^4 conidia/ml and incubated for seven days at 30C, it was mixed with rooting mixture for tomato at 10% v/v to control crown rot (*Fusarium oxysporum f.sp. radicislycopersici*). The disease was reduced by 29.7% over control and yield increased by 7% (Sivan *et al.*, 1984).

Vermiculite – Wheat bran

Trichoderma was multiplied in molasses – yeast medium for 10 days. Vermiculite (Grade 4) and milled wheat bran (250 mesh) were heated in hot air oven at 70°C for three days using metal pans. Vermiculite (100g), wheat bran (3.3 g), liquid culture (14 ml) and 0.05N HCI (17.5 ml) were mixed and packed. This was immediately used for soil application (Lewis *et al.*, 1991). Vidya (1995) applied this formulation at 250 kg/ha to mung bean and found 41 % reduction in root rot (*M. phaseolina*) and 91% increase in yield.

Alginate Pellets

Sodium alginate (20g) was dissolved in 750 ml water at 40°C on a stirring hot plate. Wheat bran ground to pass through a 0.425 mm mesh screen was placed in a glass blender container with distilled water (50g/250 ml). Kaolin can also be used in place of bran. It was autoclaved for 30 min and cooled. Fermentor biomass (16-21 g/l) was added to provide 7 x 10^6 conidia and chlamydospores. The mixture containing fungus, alginate and bran or kaolin was added dropwise into 500-ml gellant solution (0.25 M $CaCl_3$, pH 5.4). As it entered, each droplet gelled and a distinct spherical bead formed. After 20 min in the gellant, beads were separated from the solution by gentle filtration, wahsed and dried for 24 hr in a stream of air at 25°C (Lewis and Papavizas, 1986).

Other Substrates

The following substrates were also found to support the growth of *Trichoderma viride* and *T. harzianum* (Kousalya and Jeyarajan, 1988). Farm yard manure, gobar gas slurry, press mud, paddy chaff, rice bran and groundnut shell.

Equipments required for the establishment of *Trichoderma* production unit

1. Microscope (Monocular)
2. Autoclave (Vertical type) 450 x 750mm
3. Fermenter (50 lt. Capacity 6KW)
4. Laminar air flow (4'x2'x2')
5. Cooker (22lt capacity)+Gas stove
6. Poly bag sealer
7. Balance (2 Kg capacity)
8. Haemocytometer
9. Refrigerator
10. Homogeniser
11. Electronic Balance
12. Spiral Kneader
13. Packing Machine

Conclusion

The decades of laboratory research has brought the usage of *Trichoderma* spp. from lab to land. It has promoted the commercial application of *Trichoderma*. Large scale and commercial production of the same is done worldwide. It would be unrealistic to expect that this biocide would replace chemical fungicides in disease control if it has increased shelf life equivalent to that of chemical pesticides. Future research should be directed towards exploiting the avenues of *Trichoderma* spp. in disease control. Biological agents achieve their effect through a variety of mechanisms and, thus avoid the development of resistance, which often renders fungicides obsolete. Future research should give as the insight necessary to improve the genetic potential of the organisms to have a broad spectrum of action against several plant pathogens, as well as the appropriate delivery systems to achieve consistent, effective, and environmentally sound biological control in a wide range of applications.

Mass Production of *Pseudomonas fluorescens*

Introduction

Pseudomonas fluorescens (Pf-1) is a gram-negative bacterium known to positively influence plant growth, vitality and the ability of the plant to cope with pathogens resulting in higher yield. It is a potential biocontrol agent used to control the foliar and soil borne pathogens *viz., Pyriculaira oryzae, Pyriculaira grisea, Colletotrichium* sp, *Pythium* sp, *Macrophomina* sp, *Rhizoctonia* sp. and *Fusarium* sp. in dryland and irrigated crops. Apart from diseases, Pf1 reduces the pest and nematode population in agricultural and horticultural crops. *Pseudomonas fluorescens* is capable of surviving in the spermosphere, rhizosphere and phyllosphere of plants. It is found to be effective against the fungal, bacterial, viral pathogens, insect pests and nematodes. A number of strains of *P. fluorescens* suppress plant diseases by protecting the seeds and roots from fungal infection. This effect is the result of production of a number of secondary metabolites including antibiotics, siderophores and hydrogen cyanide. Competitive exclusion of pathogens as the result of rapid colonization of the rhizosphere by *P. fluorescens* may also be an important factor in disease control. It promotes the plant growth and enhances the yield potential of many crops.

Distribution

Pseudomonas spp. are common inhabitants of wide range of soil, fresh or marine water and plant surfaces (mainly rhizosphere). They are favored by moist soil with high in organic matter and are especially prevalent in rhizosphere (or) on the rhizoplane. They are capable of growing over a wide temperature range, including at near freezing temperatures and at 35-37°C, depending on the strain. Many *Pseudomonas* live in a commensal relationship with plants, utilizing nutrients exuded from plant surfaces and surviving environmental stress by occupying protected sites provided by the plant's architecture. These commensal species can have profound effects on plants by suppressing pests, enhancing access to key nutrients, altering physiological processes or degrading environmental pollutants by producing a wide variety of metabolites that are toxic to plant pathogens.

Biodiversity

Simple fluorescent pseudomonads are common inhabitants of soil and water, and were first described by Flugge (1886). Flugge recognized two biotypes, distinguishable on the character of gelatin liquefaction, which now bear the names of *Pseudomonas fluorescens* (liquefying) and *Pseudomonas putida* (non liquefying). Some other species have been recognized by virtue of their ability to synthesis, in addition to the group specific fluorescent pigment, a type-specific phenazine pigment.

From the available descriptions, some of these phenazine producers (*P. chlororaphis, P. aureofaciens, P. lemonnieri*) appear to resemble closely in most phenotypic respects the simple fluorescent pseudomonads of the type of *P. fluorescens*. However, *P. aeruginosa*, the type species of the genus *Pseudomonas*, has several properties in addition to the production of a specific phenazine pigment that set it apart from the rest of the group. It is the only fluorescent pseudomonads, has a temperature maximum (above 41°C) higher than that of the other fluorescent pseudomonads and its flagellation is invariably monotrichus, while nearly all other fluorescent strains are multitrichus. Some of the other type species of the genus *Pseudomonas* include *P. Stutzeri, P. mindocina* and *P. cepacia* (no fluorescent pigment production).

Biology

Single celled, straight or curved, 0.5-1.0 x 1.5-4.0 um in size, Gram negative, motile by polar flagella (mono/multitrichus), no sheaths or prothecae are produced, no resting stages known, metabolism respiratory, never fermentative, strict aerobes, frequently develop fluorescent, diffusible pigments of green, blue, violet, lilac, rose or yellow, particularly in iron deficient media, many species develop no pigment.

Lifecycle

In autecological studies, bacterial growth can be modeled with four different phases: lag phase (A), exponential or log phase (B), stationary phase (C), and death phase (D).

Growth is shown as L = log (numbers) where numbers is the number of colony forming units per ml, versus T (time.)

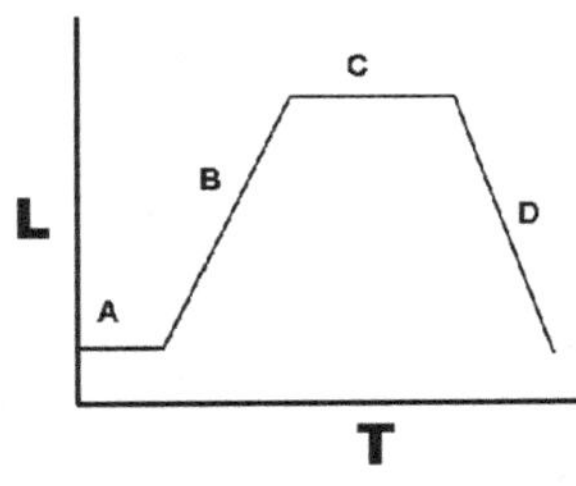

1. During lag phase, bacteria adapt themselves to growth conditions. It is the period where the individual bacteria are maturing and not yet able to divide.
2. During the exponential phase, the number of new bacteria appearing per unit time is proportional to the present population. This gives rise to the classic expponential growth curve, in which thelogrithm of the population density rises linearly with time (see figure). The actual rate of this growth (i.e. the slope of the line in the figure) depends upon the growth conditions, which affect the frequency of cell division events and the probability of both daughter cells surviving. Exponential growth cannot continue indefinitely, however, because the medium is soon depleted of nutrients.
3. During stationary phase, the growth rate slows as a result of nutrient depletion. This phase is reached as the bacteria begin to exhaust the resources that are available to them.
4. At death phase, bacteria run out of nutrients and die.

   In reality, these phases are not so well defined, and the curve is much more continuous.

Formulation Development

A successful biocontrol formulation is one, which is economical to produce, safe, stable in the environment and easily delivered and should enable the bio-control agent to act effectively and consistently under varied environmental conditions. The major factor in the formulation is the nutrient status of the substrate or additives, since the establishment of the antagonist in soil is one of the main difficulties to overcome in the application of antagonistic microorganisms. Introduction of antagonists through organic carriers alleviate the competition from autochthonous microorganisms, since the organic carriers serve as both protection and a food base during the establishment phase of the antagonist (Steinmetz and Schonbeck, 1992).

Talc Based Formulation of *Pseudomonas fluorescens* (Pf1)

Small Scale

Transfer a loopful of *P. fluorescens* culture to 100ml of sterilized Kings' B broth in a 250ml conical flask under aseptic condition and incubate on a rotary shaker (150rpm) at room temperature (28 ± 2°C) for 48hrs. Assess the population of *P. fluorescens* in the broth after the incubation period is over (A minimum of 9 x 10^9/cfu ml is necessary). Adjust pH of the substrate (talc powder) to 7 by addition of 150g calcium carbonate / Kg of talc powder and sterilize it in an autoclave at 15 psi for 30 min on two successive days. Transfer 1000g of the sterilized substrate (talc powder) and 10g of sticker, carboxy methyl cellulose into a polythene bag or any sterile container under aseptic conditions and add 400ml of *P. fluorescens* suspension. Mix thoroughly and shade dry until it reaches moisture content of less than 20 per cent. The formulation can be stored in milky white polythene bags for 3-4 months. The bioformulation should contain a minimum of 2.5 x 10^8 cfu/g.

Large Scale

The bacterium can also be multiplied in a fermentor. Prepare Kings' B broth in 500ml conical flasks and sterilized at 15psi for 30 min. Inoculate with a loopful of bacterial culture and incubate for 48hrs. In a fermentor of 250 litres capacity, prepare 200 litres of Kings' B broth and sterilize. Add the inoculum, growth in conical flasks to the fermentor at one lit/200l and incubate it for 48hr. Assess the population by drawing samples from the fermentor and it should contain a minimum of 9 ± 2 X10^9 cfu/ml. Adjust pH of the substrate (talc powder) to 7 by addition of 150g calcium carbonate / Kg of talc powder. Sterilize it in an autoclave at 15psi for 30 min for two successive days. Take 500kg of the sterilized substrate (talc powder) after mixing with 2.5kg of carboxy methyl cellulose in any sterile container under aseptic condition and add 200l of *P. fluorescens* broth culture. Mix thoroughly and shade dry until it reaches moisture content of less than 20 per cent. The formulation can be stored in polythene bags. The product should contain a minimum of 2.5X10^8cfu/g and it has a shelf life of 3-4 months under room temperature.

Liquid Based Formulation

For developing liquid formulation of *Pseudomonas*, Nutrient broth was prepared in combination with different chemical amendments to increase the survival of *Pseudomonas* cells. The chemical amendments *viz.*, trehalose at 10 mM, polyvinylpyrollidone (PVP) at 2 % and glycerol at 10 mM were added to one litre of NA and KB broth separately as per the method described by Vendan and Thangaraju (2006). One ml of log phase culture of Pf1 was inoculated individually in each broth. An uninoculated control was maintained for each broth and the flasks were incubated at room temperature. The broth cultures were analyzed for viable cell population at monthly intervals up to decline phase.

Enumerating the Viable Cell Population

The King's B broth was prepared, sterilized and 15ml of the medium was poured in sterile petriplates. The plates were kept at room temperature for 48 h. Eight equal sectors on the outside bottom of the petridish were radially marked. Four sectors were used for replications of one dilution and four for another, allowing two dilutions per plate. Serial dilution was prepared by transfer of 1 ml each of inoculum into 9 ml sterile water blanks to get 10^{-1} dilution. Similarly the dilutions were made serially upto 10^{-9}. From the dilutions, 10 µl was pipetted out and placed on the respective quadrant in the petriplate. The plates were incubated at 28 ± 1°C without any disturbance and individual colonies were counted through drop plate method (Somasegaran and Hoben, 1994).

Oil Based Formulation of *Pseudomonas fluorescens* (Pf1)

Preparation of Invert Emulsions and Evaluation of Stability

The ingredients of each phase of the invert emulsions were first mixed separately and then combined in a 50:50 per cent ratio by adding the aqueous phase onto the oil phase to obtain water in- oil formulation. The aqueous phase mainly consisted of sterile distilled water, glycerin, and in some cases water-soluble wax. The wax was first melted in sterile distilled water held at 75°C in a water bath for 10 min. Glycerin was then added to this phase. In the oil phase a plant oil (s) (source: Vegetable Oil Industries Co. Ltd., Coimbatore, India), and oil-soluble emulsifier (Tween 20) were mixed. Five types of oils were tested during emulsion preparation.

A loopful of *Pseudomonas* was inoculated into the KB broth and incubated in a rotary shaker at 150 rpm for 48 h at room temperature (28 ± 2°C). After 48 h of incubation, the broth containing 9 X 10^8 cfu/ml was used for the preparation of oil based emulsion formulation. After centrifuge at 10000 rpm at 4°C, the pellet was dissolved using sterile distilled water. The suspended pellet was first introduced into the aqueous phase of the emulsion by replacing a part (22.5%) of the sterile distilled water entering in this phase with the same quantity of pellet suspension of the Pseudomonas. The remaining sterile distilled water (22.75%) was used to melt the water-soluble wax (0.75%). The wax was, then, allowed to cool before mixing with pellet suspension and then added to the third ingredient (glycerin 4.0%). Mixing and homogenization of this phase with the oil phase of the emulsion was then carried out as described above. The concentration of cells in the aqueous phase of the emulsion was 9.8x10^9 cfu/ml prepared. The emulsion was held in screw capped glass bottles (1 l capacity) at 28 ± 1°C for the duration of the viability study (Batta, 2004).

Enumerating the Viable Colonies for Population Count

The techniques were used to assess viability over time after introduction of the cells into the emulsion. Two 100 µl samples of the emulsion containing cells (9.8 x 10^9 con cfu/ml) were taken from storage bottles at weekly intervals for 30 weeks. One sample was spread on the surface of plates containing KB medium and then incubated at 25± 1°C for 14 days. The plates were incubated at 28 ± 1 °C without any disturbance and individual colonies were counted through drop plate method (Somasegaran and Hoben, 1994).

References

Backman P.A. and Rodrigue-Kabana, R. 1975. A system for the growth and delivery of biological control agents to the soil. *Phytopathology,* 65:819-821.

Baker, K.F. and Cook, R.J. 1974. Biological control of plant pathogens. W.H. Freeman and Co., San Francisco, C.A. 433 p.

Batta, Y.A. 2004. Postharvest biological control of apple gray mold by *Trichoderma harzianum* Rifai formulated in an invert emulsion. *Crop Protection*, 23:19–26

Benhamou, N., Kloepper, J.W. and Tuzun, S., 1998. Induction of resistance against Fusarium wilt of tomato by combination of chitosan with an endophytic strain: ultrastructure and cytochemistry of the host response. *Planta*, 204 :153–168.

Connick, Jr., W.J., Lewis, J.A. and Quimby Jr., P.C. 1990. Formulation of biocontrol agents for use in Plant Pathology, New Directions in Biological control. Alternatives for Suppressing Agricultural Pests and Diseases (Eds. R.R. Baker and P.E. Dunn). Alan R. Liss, Inc. New York, pp. 345-372.

Deacon, J.W. 1988. Biocontrol of soil-borne plant pathogens with introduced inocula. *Phil. Trans. R. Soc. Lond. Ser. B.,* 318: 249-264.

Elad, Y., Chet, I. and Katan, P. 1980. *Trichoderma harzianum*. A biocontrol agent effective against *Sclerotium rolfsii* and *Rhizoctonia solani. Phytopathology*, 70: 119-121.

Elad, Y., Zuiel, Y. and Chet, I. 1986. Biological control of Macrophomina phaseolina (Tassi) Goid. by *Trichoderma harzianum, Crop Prot.*, 5:288 –292

Fravel, D.R., Marois, J.J., Lumsden, R.D. and Connick, W.J., Jr. 1985. Encapsulation of potential biocontrol agents in an alginate – clay matrix. *Phytopathology*, 75: 774-777.

Henis, Y., Ghaffar, A. and Baker, R. 1978. Integrated control of *Rhizoctonia solani* damping-off of radish. Effect of successive plantings. PCNB and *Trichoderma harzianum,* pathogen and disease. *Phytopathology,* 68: 900-907.

Jeyarajan, R and Nakkeeran, S. 1996. Exploitation of biocontrol potential of *Trichoderma* for field use In: *Current trend in life sciences vol. XXI (eds. K.Manibhushan Rao and A. Mahadevan)* Today and Tomorrow's Printers and Publishers , New Delhi, India., pp. 61-66.

Jeyarajan, R. and Nakkeeran, S. 2000. Exploitation of microorganisms and viruses as biocontrol agents for crop disease management. In: *Biocontrol potential and their exploitation in crop disease management,* (Ed, Upadhyay, *et al*) Kluwer Academic / Plenum Publishers, USA pp 95-116.

Jeyarajan, R. Ramakrishnan, G.Dinakaran, D. and Sridar R. 1994. Development of products of *Trichoderma viride* and *Bacillus subtilis* for biocontrol of root rot diseases. In: *Biotechnology in India (ed. Dwivedi) Bioved Research Society*, Allahabad, pp.25-36.

Kloepper, J. W., Leong, J., Teintze, M and Schroth, M. N. 1980. *Pseudomonas* siderophores: A mechanism explaining disease suppressive soils. *Current Microbiology,* 4: 317-320.

Kok, C.J., Hageman, P.E.J., Maas, P.W.T., Postma, J., Roozen, N.J.M. and Van Vuurde, J.W.L. 1996. Processed manure as carrier to introduce *Trichoderma harzianum* Population dynamics and biocontrol effect on *Rhizoctonia solani*. *Biocontrol Sci and Tech.*, 6: 147-161.

Kousalya, G. and Jeyarajan, R. 1988. Techniques for mass multiplication of *Trichoderma viride* Pers. Fr and *T. harzianum Rifai*. National Seminar on management of crop diseases with plant products / biological agents. 10-12 January, Agricultural college and Research Institute, Madurai. pp. 32-33.

Lewis, J.A and Papavizas, G.C. 1986. Reduced incidence of *Rhizoctonia* damping off of cotton seedlings with preparations of biocontrol fungi. *Biol. Cult. Tests*, 1: 48.

Lewis, J.A, Papavizas, G.C and Lumsden, R.D. 1991. A new formulation system for the application of biocontrol fungi to soil. *Biocontrol Sci, Technol.*, 1: 59-69.

Nakkeeran, S. and Jeyarajan, R. 1996. Exploitation of antagonistic potential of *Trichoderma* for field use. Paper presented in *National Symposium on disease of plantation crops and their management of Institute of Agriculture*, Sriniketan held during February 18-20, 1996.

Nakkeeran, S. and Renukadevi, P. 1997. Seed borne microflora of pigeonpea and their management. *Pl. Dis.Res.*, 12: 103-107.

Nakkeeran, S. and Sabitha Doraisamy. 2001. Genetic improvement of *Trichoderma viride* and assessment of its efficacy against soil borne pathogens. In National Seminar on Emerging Trends in Pests and Disease Management. Center for Plant Protection Studies, TNAU, Coimbatore p.130.

Nakkeeran, S., Kousalya Gangadharan and Renukadevi, P. 1995. Efficacy of organic amendments and antagonists on root rot and wilt of pigeonpea. National Symposium on organic farming, TNAU, Madurai. Oct. 27-28, 141-142.

Nakkeeran, S., Sankar P. and Jeyarajan, R. 1997. Standardization of storage conditions to increase the shelf life of *Trichoderma* formulations. *J. Mycol. Pl. Pathol.*, 27: 60-63.

Nakkeeran, S., Subramanian, S., Ramamurthy, R. and Sankar P. 1996. Multivatious approach in the management of panama wilt of banana. *Indian J. Mycol. Pl. Pathol.,* 26: 121.

Nandakumar, R., Babu, S., Viswanathan, R., Raguchander, T. and Samiyappan, R. 2001. Induction of systemic resistance in rice against sheath blight disease by plant growth promoting rhizobacteria. *Soil Biol. Biochem.*, 33: 603–612.

Papavizas, G.C. and Lewis, J.A. 1989. Effect of *Trichoderma* and *Gliocladium* on damping- off and blight of snapbean caused by *Sclerotium rolfsii. Plant Pathol.*, 38: 277- 286.

Papavizas, G.C., Dunn, M.T., Lewis, J.A. and Beagle-Ristaino, J. 1984. Liquid fermentation technology for experimental production of biocontrol fungi. *Phytopathology,* 74: 1171-1175.

Radja Commare, R., Nandakumar, R., Kandan, A., Suresh, S., Bharathi, M., Raguchander, T., and Samiyappan, R. 2002. *Pseudomonas fluorescens* based bioformulation for the management of sheath blight disease and leaf folder insect in rice. *Crop Protect.*, 21: 671-677.

Ranganathan, K. Sridar, R. and Jeyarajan, R. 1995. Evaluation of gypsum as a carrier in the formulation of *Trichoderma viride. J. Biol. Control.,* 9: 61-62.

Sendhilvel, V., Kavitha, K., Nakkeeran, S. Raguchander, T. and Marimuthu, T. 2006. Bacteria. *Glimpses of Plant Pathology, Text book, India*, p. 23.

Sivan, A., Elad, Y. and Chet, I. 1984. Biological control. Effect of new isolates of *Trichoderma harzianum* on *Phytihum aphanidermatum. Phytopathology,* 74: 498-501.

Somasegaran, P. and Hoben, H. J. 1994. Handbook for Rhizobia-Methods in legume-*Rhizobium* technology. Springer-Verlag, New York, USA. pp. 450 .

Steinmetz,J. and Schonbeck,F.1992. Applicability of different formulations of fungal antagonists for the control of soil-borne diseases. *Bulletin OILB-SROP*, 15: 206 – 208.

Subramanian, S. and Senthamizh, K. 2006. Effect of culture filterate of *Pseudomonas fluorescens* and *T. viride* and root exudates of *Glomus mosseae* on hatching of *Rotylenchulus reniformis. Indian Journal of Nematology*, 36:133-135.

Vendan, R. T. and Thangaraju, 2006. Development and standardization of liquid formulation for *Azospirillum* bioinoculant. *Indian Journal of Microbiology*, 46(4): 379-387.

Vidhyasekaran, P. and Muthamilan, M., 1995. Development of formulation of *Pseudomonas fluorescens* for control of chickpea wilt. *Plant Dis*., 79: 780–786.

Vidhyasekaran, P. and Muthamilan, M., 1999. Evaluation of powder formulations of *Pseudomonas fluorescens* Pf1 for control of rice sheath blight. *Biocontrol Sci. Technol*., 9 :67–74.

Whipps, J.M. 1997. Developments in the biological control of soil-borne plant pathogens. *Botanical Research*, 26:1-134.

5

Formulation and Registration Protocols for Bioagents

S.Nakkeeran, K.Eraivan Arutkani Aiyanathan and M. Muthamilan

Department of Plant Pathology, Tamil Nadu Agricultural University Coimbatore- 641 003, Tamil Nadu

Introduction

Advances in pesticide were measured exclusively by the assessment of its efficacy in management alone. But today food production has ceased to be a major concern, due to stable increase in population levels and success of innovative agricultural technologies. At this juncture, the agrochemical industry that primarily recognized the farmers as their customers has failed to acknowledge the public as an important client. Perception of the public reflects that, the negative aspects of pesticides seem to outnumber their benefits. The increased reflection on environmental concern over pesticide use has been instrumental in a large upsurge of biological disease control. Development of fungicide resistance among the pathogens, ground water and foodstuff pollution and the development of oncogenic risks has further encouraged for the exploitation of antagonistic microflora in disease management.

Biological control of plant pathogens can be broadly defined as the reduction of inoculum density or disease producing activities of a pathogen or parasite in its active or dormant state by one or more organisms, accomplished naturally or through manipulation of environment, host or antagonist or by mass introduction of one or more antagonists. The decades of laboratory experiments have led to the explosion in the field of bio-control and introduced more than 50 commercial formulations comprising of fungi and bacteria. To have a successful management of plant diseases and to have a pollution free environment, the biocontrol agents has to be ensured for its availability for commercial use by the end users. To harness the potential of biopesticides by the stakeholders and end users the product has to be registered under Central Insecticide Board authorities concerned of the respective countries.

Biopesticides Included Under the Insecticide Act, 1968

As per the Gazette Notifications No.GSR 200 (E) dated 29/03/78, GSR 839 (E) dated 14/11/85, GSR 13 (E) dated 05/01/90, GSR 621 (E) dated 27/10/97, G.S.R.224 (E) dated 26/03/99, G.S.R. 868 (E) dated 15/11/2000, G.S.R. 69 (E) dated 5/02/2001, G.S.R. 378 (E) dated June 24/06/2004, G.S.R. 198 (E) dated 31/03/2005, G.S.R. 51(E) dated 6/02/2006, G.S.R. 683(E) dated 31/10/2006 and G.S.R. 650(E) dated 11/09/2008 of the Central Insecticide Board, Dept. of Agriculture and Cooperation, Ministry of Agriculture, GOI, the following biopesticides have been included under the Insecticide Act 1968.

Registration Rules and Regulations Under Section 1968

Any person desiring to import or manufacture any insecticide may apply to the Registration Committee for the registration of such insecticide and there shall be separate application for each such insecticide.

Provided that any person engaged in the business of import or manufacture of any insecticide immediately before the commencement of this section shall make an application to the Registration Committee within a period of [seventeen months] from the date of such commencement for the registration of any insecticide which he / she has been importing or manufacturing before that date:

Provided further that where any person referred to in the preceding proviso fails to make an application under that proviso within the period specified therein, he may make such application at any time thereafter on payment of a penalty of one hundred rupees for every month or part thereof after the expiry of such period for the registration of each such insecticide.

1. Every application under sub-section (1) shall be made in such form and contain such particulars as may be prescribed.

   On receipt of any such application the registration of an insecticide, the Committee may, after such inquiry as it deems fit and after satisfying itself that the insecticide to which the application relates conforms to the claims made by the importer or by the manufacturer, as the case may be, as regards [on such conditions as may be specified by it] and on payment of such fee as may be prescribed, the insecticide, allot a registration number thereto and issue a certificate of registration in token thereof within a period of twelve months from the date of receipt of the application.

2. Provided that the Committee may, if it is unable within the said period to arrive at a decision on the basis of the materials placed before it, extend the period by a further period not exceeding six months. Provided further that if the Committee is of opinion that the precaution claimed by the application as being sufficient to ensure safety to human beings or animal are not such as can be easily observed or that notwithstanding the observance of such precautions the use of the insecticides involves serious risk to human beings or animals, it may refuse to register the insecticide.

3a. In the case of applications received by it prior to the 31st March, 1975 notwithstanding the expiry of the period specified in sub-section (3) for the disposal of such applications, it shall be lawful and shall be deemed always to have been lawful for the Registration Committee to dispose of such applications at any time after such expiry but within a period of one year from the commencement of the Insecticides (Amendment) Act, 1977 (24 of 1977):

Provided that nothing contained in this sub-section shall be deemed to make any contravention before the commencement of the Insecticides (Amendment) Act, 1977 (24 of 1977), of a condition of a certificate of registration granted before commencement, an offence punishable under this Act.

3b. Where the Registration Committee is of opinion that the Insecticide is being introduced for the first time in India, it may, pending any inquiry, register it provisionally for a period of two years on such conditions as may be specified by it.

3c. The registration Committee may, having regard to the efficacy of the insecticide and its safety to human beings and animals, vary the conditions subject to which a certificate of registration has been granted and may for that purpose require the certificate-holder by notice in writing to deliver up the certificate to it within such time as may be specified in the notice.

4. Notwithstanding anything containing in the section, where an insecticide has been registered on the application of any person, any other person desiring to import or manufacture the insecticide or engaged in the business of, import or manufacture thereof, shall on application and on payment of prescribed fee be allotted a registration number and granted a certificate of registration in respect thereof on the same conditions on which the insecticide was originally registered.

Manner of Registration

- An application for registration of an insecticide under the Act shall be made in Form I and the said Form including the verification portion, shall be signed in case of an individual by the individual himself or a person duly authorised by him ; in case of Hindu Undivided Family, by the Karta or any person duly authorised by him ; in case of partnership firm by the managing partner ; in case of a company, by any person duly authorised in that behalf by the Board of Directors; and in any other case by the person in-charge or responsible for the conduct of the business. Any change in members of Hindu Undivided Family or partners or the Board of Directors or the person in charge, as the case may be shall be forthwith intimated to the secretary, Central Insecticides Board and Registration Committee and the Licensing Officer.
- The Registration Committee may, if necessary, direct inspection of the 'testing facility' for establishing the authenticity of the data.
- An application form duly filled together with a bank draft of Rs. One hundred only, drawn in favour of the Accounts Officer, Directorate of Plant Protection, Quarantine & Storage, payable at Faridabad towards registration fee shall be sent to the Secretary, Regiatration Committee, Directorate of Plant Protection, Quarantine & Storage, NH-IV, Faridabad-121001, Haryana. One self addressed stamped envelope and one stamped envelope must be enclosed along with the application.
- The registration fee payable shall be paid by a demand draft drawn on the State Bank of India, Faridabad, in favour of the Accounts Officer, Directorate of Plant Protection, Quarantine and Storage, Faridabad, Haryana.
- The certificate of registration shall be in Form II or Form II-A, as the case may be and shall be subject to such conditions as specified therein.

Biopesticides that can be Registered with CIB

1. Botanicals

- Herbal Extract (Mixture of diallyl disulphide, allyl propyl disulphide and allyl isothiocyanate)
- Neem products (Neem leaves, Neem oil, Neem seed kernel, Neem cake)
- Karanjin (isolated from *Ponqamia glabra*)
- Extract of *Cymbopogan* Species
- Oxymatrine (plant derived pesticide- *Sophora flavescens*)

- *Triptericium wilferdii* Hook GTW - Plant extact
- Bitterbarkomycin (Plant extract of *Apocynum venetum*)
- Squamocin (Seed extract of Plant *Annona squamosa* Linn)
- *Eucalyptus* leaf extract

2. Antagonistic Fungi and Bacteria

- *Bacillus subtilis* Cohen
- *Bacillus* spp.
- *Streptomyces griseoviridis*
- *Streptomyces lydicus*
- *Gliocladium* spp.
- *Pseudomonas fluorescens* Migula
- *Trichoderma* spp.
- *Burkholderia cepacia*
- *Agrobactarium radiobacter* strain 84
- *Agrobactarium tumefaciens*
- *Erwinia amylovora* (Hairpin protein)
- *Alcaligenes* spp.
- *Ampelomyces quisqualis*
- *Candida oleophila*
- *Coniotyrium minitans*
- *Pythium oligandrum*
- *Phlebia gigantean*
- *Penicillium islanidicum*
- *Aspergillus niger*-strain AN27
- VAM-Vesicular arvescular mycorrhijal
- *Piriformospora indica*
- *Photorhabdus luminescences akhurustii* Strain K-I
- *Serratia marcescens* GPS 5

3. Entomopathogenic Fungi

- *Beauvaria bassiana* (Bals.) Vuill.
- *Metarrhizium anisopliae* (Metschnikoff) Sorokin
- *Nomuraea rileyi* (Farlow) Samson
- *Verticillium lecanii* (Zimm.)
- *Verticilium chlamydosporium*
- *Hirsutella* species
- Baculoviruses
- GVs
- NPVs

The data requirements for the registration of biocontrol agents/biopesticides were streamlined as per the recommendation of DBT. The revised guidelines have been issued by CIB in November 2000. The guidelines/data requirement for provision registration and regular registration of biopesticides are under *Section 9(3B)* and *Section 9(3)* of the Insecticide Act, 1968. List of Registered Biopesticides.

Registration Requirements of Biopesticides

Product characteristics

Formulation

A. Biological Characteristics and Chemistry

1. Systematic name (Genus and species, strain if any)
2. Common name, if any
3. Source and origin
4. Specification of the product
5. Composition of the product
6. Manufacturing process
7. An appropriate test procedure and criteria used for identification by DNA test (Restriction enzyme analysis test).
8. Method of analysis
9. Viralunit: NPVs $1x10^9$ POB/ml or g, minimu GVs: $5x10^9$ capsules/ml or g minimum

10. Contaminants
11. Shelf-life claim: Not less than 6 months
12. A sample for verification (100 ml or g)

B. Bioefficacy

Field Test

Results of two trials in two agro climatic conditions conducted by CSIR/SAU/ICAR/ICMR institutes. Data should be certified either by the university authorities or Head of the institute concerned.

Laboratory Studies

Results of two trials in two agro climatic conditions conducted by CSIR/SAU/ICAR/ICMR institutes. Data should be certified either by the university authorities or Head of the institute concerned.

C. Toxicity

a. For formulated product to be directly manufactured (Mammalian Toxicity Testing of Formulations)

i. Single dose oral (Rat and mouse) Toxicity / Infectivity /Pathogenicity
ii. Single dose pulmonary Toxicity / Infectivity /Pathogenicity (Intratracheal preferred)
iii. Single dose intravenous Toxicity / Infectivity /Pathogenicity
iv. Human safety records (Effect of lack of effects)
v. Primary skin & eye irritation
vi. Cell culture

b. For Primary culture

i. Single dose oral (rat and mouse)
ii. Single dose pulmonary
iii. Single dose intravenous
iv. Cell culture
v. Human safety records.

c. For formulation

i. Data or primary culture with technical

ii. Single dose oral (rat and mouse)

iii. Single dose pulmonary

iv. Primary skin irritation

v. Primary eye irritation.

vi. Human safety records

D. Environmental safety testing: core information requirements (for formulation only)

i. Non-target vertebrates

Birds (two species)

Fresh water fish

ii. Non-target invertebrates

- Terrestrial invertebrates
- Soil invertebrates

Information on infection and pathogenicity in mammals will be available from mammalian safety testing.

- Information on infection and pathogencity, suggested test: single-dose, oral test. *Suggested test species: bobwhite test species; rainbow trout, blue gill sunfish, Tilapia mossambica.*
- Information on infection and pathogenicity. Suggested test species; rainbow trout, blue gill sunfish, *Tilapia mossambica.*
- Information on mortality effects. It is recommended that information be obtained for honeybee and silkworm.
- Information on mortality effects. It is recommended that test species include an earthworm, *Lumbricus terrestris* or other appropriate macroinvertebrates of ecological significance.

E. Processing, packaging and labeling of Formulation

a. Manufacturing process/process of formulation:

i. Raw material

ii. Plant and machinery

iii. Unit process operation/unit process

iv. Output (Finished product & generation of waste)

b. Packaging

i. Classification-solid, liquid or other type product.

ii. U nit pack size - in metric system

iii. Specification; Details of primary, secondary & transport pack.

iv. Compatibility of primary pack with the product (Glass bottles are not recommended).

c. Labels and leaflets

As per Insecticides rules, 1971/as per existing norms indicating the common name, composition, antidote/storage statements etc. (The packaging material should also be ensured free from contamination from handling, storage and transportation).

List of Registered biopesticides based on botanicals, bacteria, fungi and viruses

Botanicals

- Azadirachtin (Neem Products)
- Pyrethrins (pyrethrum)

Antagonistic fungi and bacteria

Bacillus thuringiensis/Bacillus subtilis

- *Bt*. Var. *israelensis* 164, serotype H-14 WP formulation
- B*t* var. *kurstaki* strain A-97, serotype H3a, 35 WP
- *Bt*. Var. *israelensis* serotype H-14, strain VCRC B-17 slow release granules
- *Bt*. Var. *israelensis* serotype H-14, strain VCRC B-17 WP formulation
- *Bt*. Var. *israelensis* (WS)
- *Bt*. Var. *israelensis* serotype H-14 12(AS)

***Trichoderma* spp.**

- *Trichoderma viride* WP
- *Trichoderma viride* 1% WP

- *Trichoderma viride* 0.50% WS
- *Trichoderma harzianum* 0.50% WS
- *Trichoderma*1.15% WP*

***Pseudomonas* spp.**

- *Pseudomonas fluorescens* 1.25% WP

Entomogenous fungi

- *Verticillium lecanii*
- *Beauveria bassiana* (WP)

Baculovirus

- NPV of *Helicoverpa armigera* (AS)
- NPV of *Helicoverpa armigera* 0.43% (AS)
- NPV of *Spodoptera litura* 0.50% (AS)

Further details may be seen at http://www.cibrc.nic.in/guidelines.htm

6

Bioagents As A Component of IPM

S.Harish and S.Rajamanicam

Department of Plant Pathology, Tamil Nadu Agricultural University
Coimbatore- 641 003, Tamil Nadu

Introduction

A sustainable approach to managing pests by combining biological, cultural, physical and chemical tools in a way that minimizes economic, health and environmental risks. The objective for IPM practitioners is to minimize risks to human health and the environment from the pest management actions implemented. IPM is a holistic approach that seeks to manage pests by using methods that are effective, economically sound, and ecologically compatible. IPM practitioners base decisions on information that is collected systematically as they integrate economic, environmental, and social goals. IPM promotes the use and integration of biological control agent as one of the strategies for pest and disease management. Because, biological control of plant pathogen is a potential alternative to the use of environment harming chemical pesticides in agriculture.

Bioagents as a Component of IPM

A large number of plant diseases have been successfully controlled through bacterial antagonists. *Bacillus*-based biological control agents (BCAs) have great potential in integrated pest management (IPM) systems. BCAs are alternatives to synthetic chemical fungicides or bactericides and not as part of an integrated management system. The integration is important because the consistency and degree of disease control by *Bacillus*-based BCAs is rarely equal to the control afforded by the best fungicides or bactericides. The rice plants (cv. ADT39) applied with FZB 24 through seed treatment @ 4g/kg + seedling dip @ 4g/l + soil application @ 500g/ha + foliar application @ 500g/ha reduced t severity of be disease severity of bacterial leaf blight (Nagendran *et al.*, 2013). Raupach and Kloepper (1998), who showed that, when *Bacillus subtilis* isolate INA7 was used with *Bacillus subtilis* isolate GBO3 and *Curtobacterium flaccumfaciens* isolate ME1 as seed treatments, greater growth promotion and numerically better angular leaf spot control of cucumber was achieved where methyl bromide fumigation was used compared with that of nonfumigation. The mixtures of two *Bacillus cereus* isolates were used with plastic row mulches, control of tomato early blight was equal to the control achieved with weekly applications of chlorothalonil on tomatoes grown on bare soils (Ploper *et al.*, 1992).

Pseudomonas fluorescens is one of the most effective biocontrol agents currently available for the management of plant disease. Primarily for seed, soil and foliar treatment due to its efficient antagonistic activity against various plant pathogens. The seed treatment of rice with *P.fluorescens* (PFS) along with foliar application of salicylic acid (15 DAT) and PF2 (30 DAT) significantly reduced the rice blast severity. Also, Application of *P.fluorescens* along with salicylic acid significantly increased the disease resistance (Shyamala and Sivakumaar, 2012.). Purple blotch caused by *Alternaria porri* is a yield limiting disease of onion commonly prevailing in almost all onion growing areas of the world. The seed treatment using *Pseudomonas fluorescens* (5 g/kg) followed by two sprays of difenconazole (0.1%) interspersed with *P. fluorescens* (0.5%) spray reduced the purple blotch onion caused by *Alternaria porri*. Whereas, the seed treatment with *P. fluorescens* followed by two sprays of iprodione + carbendazim (Quintal) (0.2%) interspersed with *P. fluorescens* (0.5%) spray was on par with standard check (two sprays of difenconazole at 0.1%) (Savitha *et al.*, 2014). The combined foliar application of *Pseudomonas fluorescens* Pf1 (2.5 kg/ha) and azoxystrobin (250 ml/ha) reduced the severities downy mildew as well as powdery mildew disease of cucumber (*Cucumis sativus* L.).

Soil drenching with captan @ 0.2 per cent followed by sowing of treated seeds with *Trichoderma viride* @ 1 per cent after three days of captan drenching gave maximum seedling stand of tomato in comparison to control. In tomato, integration of imidacloprid with *Trichoderma* also gave good control of damping off and whitefly. The seed treatment with *Trichoderma* @ 1 percent or soil application of pencycuron @ 400ppm was good for chilli seedling stand. Maximum cabbage seed germination and seedling stand was observed when nursery bed was drenched with copper hydroxide@400ppm/m^2 area and seed treated with T. *viride*@ 1 per cent. Integration of *Trichoderma,* pencycuron, copper hydroxide, imidacloprid is effective and safe, hence recommended for the integrated management of major diseases of nursery in this area (Pandey *et al.*, 2004). The combination of carbendazim seed treatment 2g / kg of seeds +

soil application of *P. fluorescens, T. viride* each @ 2.5kg /ha in FYM @ 50 kg /ha reduced pigeonpea wilt caused by *Fusarium udum*. Similarly, combination of carbendazim seed treatment @ 2g/kg of seeds + ZnSo4 @ 25kg/ha soil application also reduced the pigeonpea wilt disease (Mahesh *et al*., 2010). The onion bulbs collected from the field implemented with IDM package consisting of NPK @ 100:50:50 kg ha-1 + FYM @ 10 t ha-1 + vermicompost @ 1 t ha-1 + *P. fluorescens* @ 5 kg ha-1 + copper oxychloride @ 0.3%) showed the lowest incidence of basal rot of onion caused by *Fusarium oxysporum* (Gupta and Gupta, 2014). Rajendran *et al*. (2003) reported that two isolates of *Bacillus* spp *viz*., EPCO 102 and EPCO 16 from cotton exhibited highest inhibition of *R. solani in vitro* whereas seed bacterization with EPCO 102 effectively reduced damping off disease incidence of cotton under green house conditions. The combination of *Trichoderma harzianum*, *Trichoderma asperellum* and *Trichoderma virens* reduced the cucumber root rots caused by *Fusarium solani* and it can be considered as a protective strategy in fruiting plants for reduce the negative effects of infectious pathogens (Akrami, 2015).

The black pepper three methods of treatment with biocontrol inocula *viz*. application to the nursery, preplant application and field application to the standing crop has been suggested. Fortification of solarized nursery mixture to raise the nursery stock ensured healthy and robust plants. Pre-plant application of biocontrol inoculums to the planting pit along with FYM and also field application of neem cake (1 kg/vine) mixed with 50g of inoculum to the standing crop, during May – June and subsequently one more dose of inoculum @ 50g/vine during August, did decrease the foliar yellowing and vine death. The seed pelletting with *Trichoderma* and soil application of *Trichoderma* along with neem cake @ 1.5kg/ha showed greater protection against root rot in fenugreek caused by *Rhizoctonia solani*. Besides these *T. viride, B. subtillis* and *Pseudomonas fluorescens* were also found effective in reducing coriander wilt and increase yield. Green and blue molds caused by *Penicillium digitatum* and *P. italicum*, respectively, are the main pathogens responsible for post-harvest losses of citrus fruits. The yeast isolates selected by initial screening were tested, alone and in combination with two fungicides (Thiabendazole and Imazalil), against green and blue mold, on various citrus cultivars under cold storage and laboratory (21±1°C) conditions. When the yeasts were combined with lower doses of TBZ and Ýmazalil, the disease severity caused by *P. digitatum* and *P. italicum* was reduced by 90-100% (Sharma and Amerika Singh, 2006; Vidhyasekaran, 2006). Barka *et al*. (2002) reported the *in vitro* bacterization of grapevine with plant growth promoting *Rhizobacterium*, *Pseudomonas* strain PSJN reduced the incidence of grey mould caused by *Botrytis cinerea* when compared to non bacterized control.

The chilli seeds treated with Raw cow milk for 24 h in 1 : 1 ratio (i.e. RCM diluted to 50% by adding water) at the room temperature and *T. viride* (6 g/kg seed) and *T. viride* (10 g/m^2) in nursery soil followed by dipping of nursery-raised saplings enable the chilli plant resistance against *Chilli leaf curl virus* by stimulation of defence mechanisms by metabolic changes along with increase in defence-related enzymes such as polyphenol oxidase and peroxidise (Arun Kumar, 2014). Two PGPR strains (*Pseudomonas fluorescens* FB11 and *Rhizobium leguminosarum* bv. *viceae* FBG05) isolated from roots of faba bean plants are singly and in combination as seed inoculants showed induction of systemic resistance in faba bean against *Bean yellow mosaic Potyvirus* (BYMV). The results demonstrated that BYMV challenged plants emerged from *Pseudomonas* inoculated seeds not only showed a pronounced and significant reduction in percent disease incidence (PDI) (Elbadry *et al.*, 2006). Treatment of melon seeds and soil with *P. fluorescens* suspension induced significant reduction in virus accumulation in the plants by inducing systemic resistance against CMV in the plants (Al-Ani and Adhab, 2012). Combined application of Imidacloprid (0.0375 %) on 7 DAT; *P. fluorescens* (2%) + butter milk (0.3 %) under field condition reduce the severity of the Pea nut bud Necrosis virus (PbNv) (Thiribhuvanamala *et al.*, 2013).

References

Akrami, M. 2015. Effects of *Trichoderma* spp. in Bio-controlling *Fusarium solani* and *F. oxysporum* of Cucumber (*Cucumis sativus*). *J. Appl. Environ. Biol. Sci.*, 4(3): 241-245.

Al-Ani, R.A. and Adhab M. A. 2012. Protection of melon plants against *Cucumber mosaic virus* infection using *Pseudomonas fluorescens* biofertilizer. *African Journal of Biotechnology*, 11(100):16579-16585.

Anand, T., Chandrasekaran, A., Kuttalam, S., Raguchander, T. and Samiyappan, R. 2009. Management of Cucumber (*Cucumis sativus* L.) Mildews through Azoxystrobin-Tolerant *Pseudomonas fluorescens*. *J. Agric. Sci. Technol.*, 11: 211-226.

Arun Kumar, 2014. Experimental Validation of Indigenous Knowledge for Managing Crop Diseases in Arid Rajasthan. *International Journal of Applied Life Sciences and Engineering (IJALSE)*, 1 (1) 20-27.

Barka, E.A., Gognies, S., Nowak, J., Audran, J.C. and Belarbi, A. 2002. Inhibitory effect of endophyte bacteria on *Botrytis cinerea* and its influence to promote the grapevine growth. *Biological control*, 24: 135-142.

Elbadry, M., Taha, R. M., Eldougdoug, K. A. and Gamal-Eldin, H. 2006. Induction of systemic resistance in faba bean (*Vicia faba* L.) to *Bean yellow mosaic potyvirus* (BYMV) *via* seed bacterization with plant growth promoting rhizobacteria. *Journal of Plant Diseases and Protection*, 113(6):247-251.

Gupta, R. C. and Gupta, R. P. 2013. Effect of integrated disease management packages on diseases incidence and bulb yield of onion (*Allium cepa* L.). *SAARC J. Agri.*, 11(2): 49-59.

Mahesh, M., Saifulla, M., Sreenivasa, S. And R.Shashidhar, K. 2010. Integrated management of pigeonpea wilt caused by *Fusarium udum* Butler. *EJBS*, 2 (1): 1-7.

Nagendran, K., Gandhi Karthikeyan, Faisal Peeran, M., Raveendran, M., Prabakar, K. and Raguchander, T. 2013. Management of bacterial leaf blight disease in rice with endophytic bacteria. *World Applied Sciences Journal*, 28 (12): 2229-2241.

Pandey, K.K., Pandey, P.K. and Mishra, K.K. 2004. Integrated disease management of vegetable nursery. *Veg. Sci.*, 31 (1): 45-50.

Ploper, L.D., Backman, P. A., Stevens, C., Khan, V. A., and Rodriquez- Kabana, R. 1992. Effects of soil mulch, row cover, and biological and chemical foliar treatments on early blight of tomato. *Biol. Cult. Tests*, 7: 38.

Rajendran, L., Harish, S., Radjacommare, R., Kandan, A., Sible, G.V., Lavanya, N., Durairaj, C., Ramanathan, A. and Samiyappan, R. 2003. Bacterial endophytes as potential biocontrol agents for important pest and disease of cotton. Proceedings on the 6th International Plant Growth Promoting Rhizobacteria (PGPR) Workshop held at Calicut, Kerala, India, 5-10 October, 2003. Abstracts and Short Papers. Pp.202-207.

Raupach, G.S., and Kloepper, J.W. 1998. Mixtures of plant growthpromoting rhizobacteria enhance biological control of multiple cucumber pathogens. *Phytopathology*, 88: 1158-1164.

Savitha, A. S., Ajithkumar, K. and Ramesh. G. 2014.Integrated disease management of purple blotch *[Alternaria porri* (Ellis) Cif] of onion. *Pest Management in Horticultural Ecosystems*, 20(1): 97-99.

Sharma P. and Amerika Singh. 2006. Strategies for integrating antagonistic organisms in the crop disease management programme in India. 144-152.

Shyamala, L. and Sivakumaar, P.K. 2012. Integrated control of blast disease of rice using the antagonistic rhizobacteria *Pseudomonas fluorescens* and the resistance inducing chemical salicylic acid. *International Journal of Research in Pure and Applied Microbiology*, 2(4): 59-63.

Thiribhuvanamala, G., Murugan, M., Jayalakshmi, V., Manoranjitham, S.K. Renukadevi, P. and Rabindran, R. 2013. Strategic approaches for the management of pea nut bud necrosis virus disease of tomato. *Pest Management in Horticultural Ecosystems*, 19(1): 67-72.

Vidhyasekaran, P. 2006. Antagonistic organisms as plant activators in current status of biological control of Plant Diseases (Etd). Ramanujam and Rabindra. BDBC (ICAR): 97-108.

7

Notes for Practical Sessions Preparation of Media for Isolation of Bioagents

S. Nakkeeran and C.Dheepa

Deprtment of Plant Pathology, Tamil Nadu Agricultural University Coimbatore- 641 003, Tamil Nadu

To study the characteristics of microorganisms, it is essential to culture and growth them. Growth is the increase in cell number which can be obtained by providing proper physical and chemical conditions. Medium is a substance that provides nutrients for growth and multiplication of microorganisms. Any medium for the cultivation of bacteria and fungi must provide certain basic nutritional requirements which include a carbon, nitrogen, phosphate and various mineral nutrients. The following is the medium composition for the isolation and culturing of different biocontrol agents

1. Potato dextrose agar medium (Ainsworth, 1961)

It is a common medium for isolation and maintenance of wide variety of microorganisms.

Potato (peeled and sliced)	: 200 g
Dextrose	: 20 g
Agar agar	: 20 g
Distilled water	: 1000 ml
pH	: 6.0 to 6.5

2. Selective medium (King'B medium) for Pseudomonads (King *et al.*, 1954)

Peptone	: 20 g
Glycerol	: 10 ml
Magnesium sulphate	: 0.150g
Dipotassium hydrogen phosphate	: 0.150g
Agar	: 20 g
Distilled water	: 1000 ml

3. Nutrient Agar medium for Bacillus

Peptone	: 5 g
Beef extract	: 3 g
NaCl	: 5 g
Glucose	: 5 g
Agar	: 20 g
Distilled water	: 1000 ml

4. Trichoderma selective medium (TSM) for *Trichoderma*

$MgSo_4$	: 0.2 g
K_2HPO_4	: 0.9 g
Ammonium nitrate	: 1 g
KCL	: 0.15 g
Dexon 50 WP or Apron	: 0.3 g
PCNB	: 0.2 g
Rose Bengal	: 0.15 g
Chloromphenicol	: 0.25 g
Agar	: 15 g
Distilled water	: 1000 ml

5. Starch Casein medium for Actinomycetes

Soluble carbon	: 10 g g
Casein	: 0.3 g
KNO_3	: 2 g
Magnesium sulphate	: 0.05 g
Calcium carbonate	: 0.03 g
$FeSO_4$	: 0.01 g
KH_2PO_4	: 2 g
Agar	: 20 g
Distilled water	: 1000 ml

6. KenKnights Agar Medium for Actinomycetes

Dextrose	: 1 g
KH_2PO_4	: 0.1 g
$NaNO_3$	: 0.1 g
KCl	: 0.1 g
Magnesium sulphate	: 0.1 g
Agar	: 20 g
Distilled water	: 1000 ml

7. Yeast Peptone Extract (YPD) for Yeast (*Saccharomyces, Metschnikowia, Candida*)

Yeast Extract	: 10 g
Dextrose	: 20 g
Peptone	: 20 g
Agar	: 20 g
Distilled water	: 1000 ml

8

Isolation of Antagonistic *Trichoderma*, Rhizobacteria and Actinomycetes

A.S. Krishnamoorthy and R. Priyanka

Department of Plant Pathology, Tamil Nadu Agricultural University Coimbatore- 641 003, Tamil Nadu

The isolation of microbes is commonly done using serial dilution technique.

1. *Trichoderma*

Trichoderma is one of the important biocontrol agents. The colonies first produce mycelial growth and then turn yellow and finally produce a dark green color mat.

Isolation of *Trichoderma*

- Take one gram of the rhizosphere soil sample (soil particles adhering to root system) and mix it well in 10 ml of sterile distilled water in a clean and sterilized test tube (1:10 dilution).
- Take 1 ml of the suspension from 1:10 dilution and add to nine ml of sterile distilled water in a tube (1:100 dilution) and make the dilution till $1:10^4$ and $1:10^5$.
- Transfer one ml of this suspension to a sterile Petri dish and 15 ml of melted and cooled *Trichoderma* selective medium.
- Rotate the plates gently and allow it to cool. Incubate the Petri dishes at room temperature for 7 days to observe the colonies.
- Express the population count of Trichoderma in CFU per gram of soil.
- Pick out and transfer pure colonies onto TSM plates/slants for culture maintenance and further studies.

2. Rhizobacteria

Pseudomonas and *Bacillus* are studied as a plant growth promoting rhizobacteria (PGPR) and effective biocontrol agent against several plant diseases. It is a gram negative, rod shaped, non spore forming bacteria.

2.1. Isolation of *Pseudomonas*

- Take one gram of the rhizosphere soil sample (soil particles adhering to root system) and mix it well in 10 ml of sterile distilled water in a clean and sterilized test tube (1:10 dilution).
- Take 1 ml of the suspension from 1:10 dilution and add to nine ml of sterile distilled water in a tube (1:100 dilution).
- Make six, seven and more serial dilutions to get $1{:}10^8$, $1{:}10^9$ and 10^{10} dilutions respectively.
- Transfer one ml of this suspension to a sterile Petri dish and 15 ml of melted and cooled King'B medium.
- Rotate the plates gently and allow it to cool. Incubate the Petri dishes at room temperature for 2 to 3 days and observed the colonies.
- Express the population count of *Pseudomonas* in CFU per gram of soil.
- Pick out and transfer pure colonies onto King's B plates / slants for culture maintenance and further studies.

2.2. Isolation of *Bacillus*

- Take one gram of the rhizosphere soil sample (soil particles adhering to root system) and mix it well in 10 ml of sterile distilled water in a clean and sterilized test tube (1:10 dilution).
- Take 1 ml of the suspension from 1:10 dilution and add to nine ml of sterile distilled water in a tube (1:100 dilution).
- Make six, seven and more serial dilutions to get $1{:}10^8$, $1{:}10^9$ and 10^{10} dilutions, respectively.
- Transfer one ml of this suspension to a sterile Petri dish and 15 ml of melted and cooled Nutrient Agar (NA) medium.
- Rotate the plates gently and allow it to cool. Incubate the Petri dishes at room temperature for 2 to 3 days and observed the colonies.
- Express the population count of *Bacillus* in CFU per gram of soil.

- Pick out and transfer pure colonies onto NA plates / slants for culture maintenance and further studies.

3. Isolation of Actinomycetes

Actinomycetes are the organisms with characteristics common to both bacteria and fungi but yet possessing distinctive features to delimit them into a distinct category. In the strict taxonomic sense, actinomycetes are clubbed with bacteria the same class of Schizomycetes and confined to the order Actinomycetales.

Procedure

- Take one gram of the rhizosphere soil sample (soil particles adhering to root system) and mix it well in 10 ml of sterile distilled water in a clean and sterilized test tube (1:10 dilution).
- Take 1 ml of the suspension from 1:10 dilution and add to nine ml of sterile distilled water in a tube (1:100 dilution) and make the dilution till $1:10^4$ and $1:10^5$

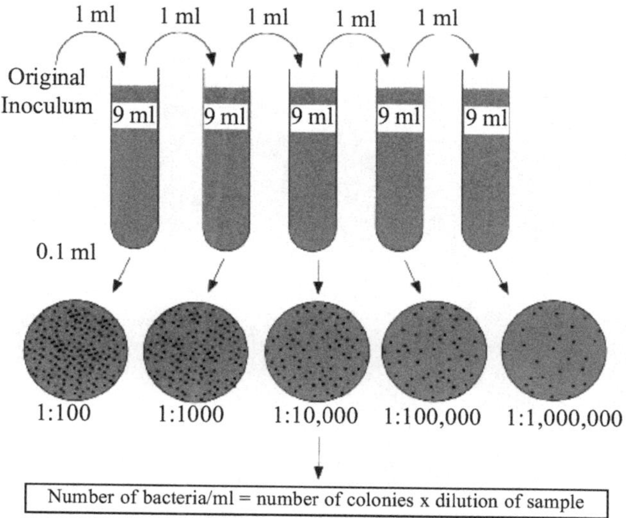

- Transfer one ml of this suspension to a sterile Petri dish and 15 ml of melted and cooled Starch Casein medium.
- Rotate the plates gently and allow it to cool. Incubate the Petri dishes at room temperature for 7 to 14 days and observed the colonies.
- Express the population count of Actinomycetes in CFU per gram of soil.
- Pick out and transfer pure colonies onto plates/slants prepared with Starch Casein medium for culture maintenance and further studies.

9

Mass Multiplication of *Trichoderma viride*

S. Nakkeeran, S.Vasumadhi and S.Vinod Kumar

Department of Plant Pathology, Tamil Nadu Agricultural University Coimbatore- 641 003, Tamil Nadu

The discovery and use of synthetic chemicals has contributed greatly to the increase of food production industry by controlling pest and diseases. However, the use of synthetic chemicals has raised number of ecological problems apart from increasing the cost of cultivation. In recent years, scientists have directed their attention in exploring the potential of beneficial microbes for plant protection measures. Biocontrol agents exhibit antagonistic activity against plant pathogens and activate resistance mechanisms in the host despite improving plant growth. In agricultural soils the nutritional condition is depleted hence beneficial microbes decrease and increase the pathogenic microbes. To ensure the sustained availability of biocontrol agents mass production technique and formulation development protocols has to be standardized which also increases the shelf life of the formulation. Though these agents have a very good potential in the management of Pest and Diseases it can not be used as cell suspension under field conditions. The cell suspensions of these biocontrol agents should be immobilized in certain carriers and should be prepared as formulations. The organic carriers used for formulation development include peat, turf, talc, lignite, kaolinite, pyrophylite, zeolite, alginate, pressmud, sawdust, vermiculate, etc.

Characteristics of an ideal formulation

- It should have increased shelf life
- It should not be phytotoxic to the crop plants
- It should dissolve well in water and should release the bacteria

- It should tolerate adverse environmental conditions
- It should be compatible with other agrochemicals
- It should protect the biocontrol agents from desiccation and death of cells

Trichoderma is the most widely used biocontrol agent. Among the various isolates of *Trichoderma, T.viride, T.harzianum, T.virens* and *T.hamatum* are used against the management of various diseases of crop plants. *Trichoderma viride* is an important *Trichoderma* species which is a potential biofungicide exploited to manage many of the soil and seed-borne plant pathogens. It is found on well decomposed organic matter, root surfaces of various plant and decaying bark. It possesses innate resistance to most agricultural chemicals which includes fungicides. The mechanism of biological control includes mycoparasitism, antibiosis, competition for nutrients and space, induced resistane, inactivation of pathogen's enzymes and making the crop tolerance to stress through enhanced root and plant development.

Molasses Yeast Medium

Yeast extract : 5g

Molasses: 30g;

Water: 1000 ml

Mass multiplication of *T. viride*

Nucleus culture

- The nucleus culture of *Trichoderma viride* (Department of Plant Pathology, Tamil Nadu Agricultural University, Coimbatore) is maintained in potato dextrose agar slants.
- *T. viride* is multiplied on potato dextrose agar in 9 cm diameter petriplates and the culture is used for subsequent inoculation in 100 ml of molasses's-yeast medium in each of 250 ml conical flasks distributed and autoclaved at 15 lbs/sq. inch pressure (121°C) for 1 h.
- Eight mm mycelial disc of *Trichoderma viride* from the petridish is taken and transferred to each flask that has been incubated for 10 days.
- This mother culture is used for inoculating medium in fermenter.

Mass multiplication

- Prepare Molasses-yeast medium (50 lt.) in 50 lt. fermenter and sterilize at 121°C, 15 lb/sq. inch. and cool it to room temperature.

- Add the inoculums of 2.5 lt. (mother culture) grown in conical flasks aseptically into the sterilized medium to the fermenter and incubate for 10 days.
- Draw samples at 10 days after incubation and assess the spore load in the broth.
- The spore load should be at least 10^8 spores / ml at the time of sampling.
- Presence of greenish fungal growth indicates *Trichoderma viride* and this broth culture contains fungal mat, spores and antibiotics.
- Mix 1 lt broth culture with 2 kg talc powder and carboxy methyl cellulose @ 5g/kg.
- Talc based *Trichoderma viride* formulation is ready for use after air drying and sieving (moisture content 12%).
- Pack the formulation in opaque polythene bags and store in a cool dry place before use.
- The product can be safely stored for four months without losing its viability.

Methods of application

Biocontrol agents can be applied to various plant parts like seed, root, foliage, inflorescens and also the soil. For this purpose several methods have been developed effectively and enhanced the yield considerably.

Seed treatment

Seed treatment is an alternative method for introducing bio-control agents into soil plant environment. The efficacy of seed treatment depends on the ability of the antagonist to multiply in the rhizosphere region. Seed treatment with *T.harzianum* reduced the disease caused by *R. solani* on cotton in field conditions (Elad *et al.*, 1982). Treatment of black gram, green gram, pigeon-pea, cowpea, groundnut, sunflower, sesamum and cotton with *T.viride* @ 4g/Kg reduced the incidence of root rot and wilt (Nakkeeran and Renukadevi, 1997).

Solid Matrix Priming

It is a process in which moistened seeds are mixed with an organic carrier and the moisture content of the mixture is brought to a level just below that required for seed sprouting. Solid matrix priming can further enhance the efficacy of antagonist in treated seeds. *Trichoderma* grow on seed surface during priming process and increase in numbers and other microbes may not easily dislodge it. Slurry of *Trichoderma* is first added to seeds followed by addition of lignite and water. *Trichoderma* required less moisture for growth. The primed seeds were incubated for 4 days at 80-100% RH. During this period *Trichoderma* colonises seed and sporulation was evident. In *Pythium ultimum* infested soil the stand was increased to 70-80% as against 10% in control. In tomato seed the population of *Trichoderma* increased by ten fold due to solid matrix priming (Harman *et al.*, 1989).

Soil

For effective management of soil borne disease, a high population of the antagonist in the soil is necessary. Adams (1990) defined effeciency of biocontrol agents as the ratio of number of propagules of mycoparasites required to obtain disease control to the typical inoculum density of a plant pathogen. Therefore attempts were made to develop suitable delivery system to soil. For controlling *R. solani*, *Trichoderma* population of 5 x 10^6 was required for each propagule of *R. solani*. Addition of wheat bran based inoculum to soil gave 80 per cent reduction of root rot over control in chickpea and bean (Elad *et al.*, 1986). Incidence of urdbean root rot was reduced by 91.3 per cent by adding *T.viride* + *T. harzianum* to soil (Kousalya and Jeyarajan, 1988). Delivering *T. harzianum* through soil, during sowing increased the per centage of survival of peanut (90%), while in control none of the plants survived (Muthamilan and Jeyarajan, 1996).

Cut Stump Application

Ricard, (1981) developed a special shears to apply *T. viride* simultaneously with prunning to protect fruit trees against silver leaf disease caused by *Chondrostereum purpureum*. The shears have a reservoir for suspension of *T. viride*. Each prunning cut was covered by spore suspension.

Hive insert

Thomson *et al.* (1992) developed an innovative method of application of bio-control agent right in the infection court at the exact time of susceptibility. *Botrytis cinerea* (fruit rot of straw berry) was effectively controlled by *T. virens* through honey bees which acted as vector. Peng *et al.* (1992) developed a novel and more efficient method of applying inoculum of *T. virens* on talc and corn meal (5 x 10^8 cfug/g), which was placed inside the beehive in a dispensor. Bees acquired the conidia on their legs and bodies as they crawled. Each bee had a spore load of 88 to 1800 x 10^3. Each flower got 1.6 to 27 x 10^3 conidia of antagonist. It suppressed *B. cinerea* on stamens (54%) and petals (47%) and controlled fruit rot.

10

Mass Multiplication of *Pseudomonas fluorescens*

S. Nakkeeran, R. Dheepa and A. Ramanathan

Department of Plant Pathology, Tamil Nadu Agricultural University Coimbatore- 641 003, Tamil Nadu

Bacteria are by far the common type of soil microorganisms, possibly because they can grow rapidly and have the ability to utilize a wide range of substances as either carbon or nitrogen sources. The bacteria that provide some benefit to plants are found near, on (or) even within the roots of plants (Kloepper *et al.*, 1980) to protect the plants from pathogen attack called as antagonistic organism. Among the many potential bacterial antagonists associated with the plant roots the fluorescent pseudomonads have received prominent attention due to their abundance in plant rhizosphere and their ability to colonize roots of a wide range of crop plants. *Pseudomonas fluorescens* is a Gram negative, non spore forming, rod shaped bacteria producing pigments which are greenish, fluorescent and water soluble. It is found to be effective against the following fungal pathogens *viz.*, *Rhizoctonia solani*, *Macrophomina phaseolina*, *Fusarium udum*, *Pyricularia grisea*, *Fusarium oxysporum* fsp. *cubense*. It is capable of surviving in the spermosphere, rhizosphere and phyllosphere of plants. The mechanisms of biological action exhibited inovles antibiosis, competition, lysis and induced resistance. *Pseudomonas* have an exceptional capacity to produce a wide variety of metabolites, including antibiotics, that are toxic to plant pathogens.

1. Raw Materials

a. Culture of *Pseudomonas fluorescens* Pf1

b. Talc powder - 300 mesh

c. Selective medium (King'B medium) for Pseudomonads

Peptone	: 20 g
Glycerol	: 10 ml
Magnesium sulphate	: 0.150g
Dipotassium hydrogen phosphate	: 0.150g
Agar	: 20 g
Distilled water	: 1000 ml

2. Requirements

a. Autoclave

b. Inoculation chamber

c. Walk-in incubation rooms

d. Shaker

e. Bottles

f. Microscope

g. Homogenizer

h. Fermenter

3. Mass Multiplication

- Inoculate loopful of *P. fluorescens* into the King's B broth and incubate in a rotary shaker at 150 rpm for 72 h at room temperature (28 ± 2°C).
- After 72 h of incubation, the broth containing 9x 10^8 CFU/ml should be used for the preparation of talc-based formulation.
- To the 400 ml of bacterial suspension, 1 kg of the purified talc powder (sterilized at 105°C for 12 h), calcium carbonate 15 g (to adjust the pH to neutral) and carboxymethyl cellulose (CMC) 10 g (adhesive) to be mixed under sterile conditions.
- The product should be shade dried to reduce the moisture content below 20 per cent and pack in polypropylene bag with appropriate seal.
- At the time of application, the population of bacterium in talc formulation was checked to 2.5 to 3 x 10^8 cfu/g.

4. Avoiding Contamination

This is necessary, while multiplying the bacteria in agar medium in Petri dish and in liquid medium in flasks. Inoculation should be done under aseptic condition using Laminar Flow Chamber whose face is located 10 feet away from the wall. Mixing of the both with the talc powder, drying and packing should be done in dust free room. Packing should be done only after thoroughly drying to 15% moisture content.

5. Safety During Production

Any contaminated flask or Petri dish should be sterilized in the autoclave and disposed.

Method of Application (*Pseudomonas fluorescens* – TNAU - Pf-1)

Field Crops

I. Paddy – Blast and sheath blight

Seed treatment: Mix paddy seeds with the formulation @10 g/kg of seed and soak the seed in water for overnight. Decant the excess water and allow the seed to sprout for 24 hours and sow the sprouted seed.

Seedling root dipping: Apply 2.5 kg of the formulation to the water stagnated in an area of 25 sq. m. The seedlings after pulling out from the nursery should be placed in stagnated water containing the bacteria. A minimum period of 30 minutes is necessary for soaking the roots and prolonged soaking will enhance the efficiency.

Soil application: Apply the product @ 2.5 Kg/ha mixed with 50 Kgs of well decomposed farmyard manure (FYM) or sand at 30 days after transplanting. Foliar application: Spray the product @ 0.2% concentration (1 Kg/ha) commencing from 45 days after transplanting at 10 days interval for 3 times depending on the disease intensity. If there is no disease incidence, a single spray is sufficient.

II. Ragi - Blast

Seed treatment: 10 g/ Kg of seed

Foliar spray: Spray the product @ 0.2 % concentration (1 kg/ha) after 50 % of the flower emergence and again repeat the spray at 10 days after the first spray.

III. Groundnut, Gingelly, Sunflower, Redgram, Greengram, Blackgram – Root rot and wilt diseases

Seed treatment: 10 g / Kg of seed

Soil application: Apply 2.5Kg / ha formulation mixed with 50 Kg of well decomposed FYM or sand at 30 days after sowing.

IV. Cotton – Root rot and wilt

Seed treatment: 10 g/ Kg of seed

Soil application: Apply 2.5 Kg/ha formulation mixed 50 Kg of well decomposed FYM / sand at 30 days after sowing.

Horticultural Crops

I. Vegetable crops – Damping off

Seed treatment: The seeds of tomato, chillies, bitter gourd, ash gourd, water melon and musk melon have to be treated with formulation at the rate of 10 g per kg of seed 24 hours before sowing to prevent the damping off disease.

Main field application: Apply talc based formulation of *Pseudomonas fluorescens* as soil application at the rate of 2.5 kg/ha mixed with well decomposed FYM 50 Kgs.

II Cabbage and Cauliflower club root disease

Seed treatment: 10 g/kg of seed

Soil application: 2.5 kg/ha *Pseudomonas fluorescens* mixed with 50 Kgs of FYM and then applied to the soil before planting.

Seedling root dip: Seedling dip in solution containing formulation of 5 g/litre of water for 30 min.

III Mango - anthracnose

Foliar spray with *Pseudomonas fluorescens* talc formulation at 0.5 % concentration with 15 days interval after fruit setting.

IV Banana - Wilt and anthracnose disease

Sucker treatment: 10 g/sucker (Paring and pralinage)

Soil Application: 2.5 Kg/ha of formulation + 50 kg of FYM or Sand. Apply once at the time of planting and repeat it once in 3 months.

Capsule Application: 50 mg/Capsule/Sucker apply once in three months from 3rd month after planting.

Bunch spray: spray the formulation @ 0.5% concentration at last hand emergence and then at 30 days interval. Combination of all the above treatments will give effective management against the wilt and anthracnose disease with increase in yield.

Uses

- TNAU-Pf-1 promotes plant growth and enhances the yield potential of many crops.
- It suppresses the soil borne pathogens like *Fusarium, Rhizoctonia, Phytophthora and Pythium* species and many nematodes.

It induces systemic resistance in plant system against plant pathogens.

11

Mass Multiplication of *Bacillus subtilis*

S. Harish, S. Rajamanicam and S. Rageshwari

Department of Plant Pathology, Tamil Nadu Agricultural University
Coimbatore - 641 003, Tamil Nadu

1. Raw Materials

a. Culture of *Bacillus subtilis*

b. Talc powder - 300 mesh

Nutrient Agar Broth (NB)

Peptone	: 5.0 g
Beef extract	: 3.0 g
NaCl	: 5.0 g
Distilled water	: 1000 ml

2. Requirements

a. Autoclave
b. Inoculation chamber
c. Walk-in incubation rooms
d. Shaker
e. Bottles
f. Microscope
g. Homogenizer
h. Fermenter

3. Mass Multiplication

- Inoculate loopful of *Bacillus* into the Nutrient broth and incubate in a rotary shaker at 150 rpm for 72 h at room temperature (28 ± 2°C).
- After 72 h of incubation, the broth containing 9x 10^8 CFU/ml should be used for the preparation of talc-based formulation.
- To the 400 ml of bacterial suspension, 1 kg of the purified talc powder (sterilized at 105°C for 12 h), calcium carbonate 15 g (to adjust the pH to neutral) and carboxymethyl cellulose (CMC) 10 g (adhesive) to be mixed under sterile conditions.
- The product should be shade dried to reduce the moisture content below 20 per cent and pack in polypropylene bag with appropriate seal.
- At the time of application, the population of bacterium in talc formulation was checked to 2.5 to 3 x 10^8 cfu/g.

4. Avoiding Contamination

This is necessary, while multiplying the bacteria in agar medium in Petri dish and in liquid medium in flasks. Inoculation should be done under aseptic condition using Laminar Flow Chamber whose face is located 10 feet away from the wall. Mixing of the both with the talc powder, drying and packing should be done in dust free room. Packing should be done only after thoroughly drying to 15% moisture content.

12

Assessing the Shelf life of Bioformulations

S. Harish, S. Rajamanicam and S. Rageshwari

Dept. of Plant Pathology, Tamil Nadu Agricultural University
Coimbatore- 641 003, Tamil Nadu

Shelf life for *Pseudomonas fluorescens* (TNAU Pf-1)

Aim: To find out the shelf life of *P. fluorescens* (TNAU Pf-1) using different carrier materials

Product: The talc based *P. fluorescens* (TNAU Pf-1) product is used for the study on shelf life.

Selective medium (King'B medium) for Pseudomonads

Peptone	: 20 g
Glycerol	: 10 ml
Magnesium sulphate	: 0.150g
Dipotassium hydrogen phosphate	: 0.150g
Agar	: 20 g
Distilled water	: 1000 ml

Protocol

- The samples derived from one month intervals/fortnight intervals starting from the zero day of formulation development should be used for the assessment of bacterial cells.
- Mix one gram of the talc product in 10 ml of sterilized distilled water in a clear and sterilized test tube (1:10).

- Take one ml of the suspension and add to nine ml of sterile distilled water in a tube (1:100). Similarly, make the dilutions to get $1:10^8$, $1:10^9$ and 10^{10} dilutions respectively.
- Transfer one ml of this suspension to a sterile Petri dish and add 15 ml of melted and cooled King'B medium.
- Rotate the plates gently and allow it to cool.
- Incubate the Petri dishes at room temperature for 2 to 3 days.
- Observe the development of typical *Pseudomonas* and calculate the number of colony units per gram of the product.
- Five samples should be derived at a single time and the analyses to be done.
- The meteorological parameters have to be recorded during the analyses period.

Determination for Contamination

Take a gram of the product and mix in 10 ml of the sterile water in a clean and sterilized test tube. Shake well and add one ml of the suspension to the nine ml of sterile water in a tube, make serial dilutions upto $1x10^9$. Transfer 1 ml of this dilution to 6 Petri dishes. Add the following cooled selective media separately to each plate. Rotate the plate gently and allow it to cool. Incubate the Petri dishes at room temperature for 2 to 3 days.

Selective Media	Test Organism
DCLS agar	Salmonella, Shigella, Vibrio
Potato Dextrose Agar	Fungal contaminants
Nutrient Agar	Bacterial contaminants

Similarly, the shelf life data for *Trichoderma viride* (using TSM) and *Bacillus subtilis* (using NA medium) can be generated.

13

Extraction of DNA from the Biocontrol Agents

L.Rajendran and S.Rageshwari

Department of Plant Pathology, Tamil Nadu Agricultural University
Coimbatore - 641 003, Tamil Nadu

Deoxyribonucleic acid (DNA) carries the genetic instruction for the biological development and functioning of all cellular forms of life. It is the blue print which contains the instructions to construct other cell components. Molecular biology experiments often require isolation of DNA for a variety of purposes like construction of DNA libraries, gene cloning, PCR and other marker technology, etc. The extraction of DNA from actively growing tissue is relatively easier yielding higher quantities of DNA.

Principle

The cell wall is ruptured and the proteins associated with DNA are removed by detergent like SDS or by interaction with phenol. Lipids are extracted with alcohol and DNA is precipitated by isopropanol.

Materials Required

Maize seed samples, Liquid nitrogen, CTAB (Cetyl trimethyl ammonium bromide), Chloroform, isoamyl alcohol, Sodium chloride, Glacial acetic acid, Isopropyl alcohol and 70% ice cold ethanol, TE buffer (5X) (24.2 g of Tris base, 5.71 ml of galcial acetic acid, 10 ml of 0.5M EDTA (pH 8.0) , dissolved in 1000 ml of distilled water). Prepare 1X TE as per the requirement.

Procedure

- One hundred milligrams of maize seed sample was taken and ground in a pestle and mortar using 1-2 ml of CTAB extraction buffer.
- After maceration 0.7 ml of ground sap was transferred into 1.5 ml of microfuge tube and incubated at 65°C for 10 min.
- After incubation, 750 µl of chloroform: isoamyl alcohol (24:1) was added and mixed to form an emulsion by inverting the tube. The mixture was centrifuged at 10000 rpm for 10 min.
- The upper aqueous phase was collected in a new 1.5 ml microfuge tube and equal volume of chloroform: isoamyl alcohol (24:1) was added, mixed and centrifuged at 10000 rpm for 10 min.
- Three hundred micro litres of aqueous phase was taken (taking care not to disturb the inter phase) and mixed with 0.5 volumes of 5M NaCl and 2 volumes of ice cold ethanol.
- It was mixed well and incubated at -20°C for 10 min and then centrifuged at 13000 rpm at 4°C for 10 min. and ethanol was decanted and pellet was dried.

Dried pellet was resuspended in 50 µl sterile distilled water and used for PCR reactions. The final concentrations of DNA were tested by 0.8 per cent agarose gel electrophoresis.

14

PCR Detection of Antibiotics Producing Genes from Antagonistic *Bacillus*

S. Harish, S. Rajamanicam and S. Rageshwari

Departmnet. of Plant Pathology, Tamil Nadu Agricultural University Coimbatore- 641 003, Tamil Nadu

Many *Bacillus* species are capable of producing a wide variety of secondary metabolites that are diverse in structure and function. The production of metabolites with antimicrobial activity is one determinant of their ability to control plant diseases. The activity is determined using PCR. PCR consists of an exponential amplification of a DNA fragment, and its principle is based on the mechanism of DNA replication *in vivo*. dsDNA is denatured to ssDNA, duplicated, and this process is repeated along the reaction according to the following formula:

$C = C_0(1+E)^n$

$E = 10^{-1/s} - 1$; if E= 1 then s= -3.3219

where,

C: final amount of DNA

$\mathbf{C_0}$: initial amount of DNA

E: efficiency

n: number of cycles

s: slope of the exponential phase.

Aim: To detect the antibiotics producing genes from *Bacillus* through PCR technique

Materials Required

For isolation: NA Broth, Bacterial culture, TE buffer, Ethanol, Lysozyme, CTAB, Chloroform : Phenol (1:1) and Icecold Propanol

For gel electrophoresis: Agarose, 1% TAE buffer and Ethdium Bromide

Buffer composition

Tris Base	: 25.2 g
Glacial Acetic Acid	: 5.71 ml
EDTA	: 10 ml
pH	: 8.0

CTAB buffer

CTAB	: 2 g
Tris HCL	: 10 ml
NACL	: 8.18 g
EDTA (0.5 M)	: 4 ml
Polyvinyl pyrolidine	: 1 g
Mercapto ethanol	: 1 ml
Sodium sulphate	: 1 g

Procedure

Isolation of DNA from *Bacillus*

- Inoculate the bacteria in NA broth (1 ml broth in eppendorf tube).
- After 18-24 h of inoculation, centrifuge the contents at 6000 rpm for 10 min at 10°C.
- Discard the supernatant and suspend the pellet with TE buffer and add 0.5 ml butanol/butyl alcohol.
- Centrifuge the contents at 6000 rpm for 10 min at 10°C and discard the supernatant.
- Add 1 ml of TE buffer to the pellet to trace out the ethanol.
- Centrifuge the contents at 6000 rpm for 10 min at 10°C and discard the supernatant.

- Add 1 ml of TE buffer with 100 ml of freshly prepared lysozyme.
- Keep it at room temperature for 5 min (to activate the lysozyme).
- Add 200 ml of 5M NaCl+150 ml of CTAB to the pellet and keep it on water bath for 10 min at 65°C.
- Add 1 ml of 1:1 Chloroform:Phenol solution and shake well.
- Centrifuge the contents at 20000 RPM for 10 min.
- Take the supernatant and transfer it into clean, fresh eppendorf tube.
- Add 0.6 ml of icecold propanol and incubate at -20°C overnight.
- Centrifuge the contents at 12000 RPM for 15 min and discard the supernatant.
- Dry the pellet for 2 hrs, dilute the pellet with 50 ml of TE buffer or add sterile water
- Add 1 ml of RNAase and incubate at -20C.
- Check the DNA quality by running the gel electrophoresis
- Take 0.4 g of agarose in 50 mo of 1% TAE buffer. Heat it for homogenizing and allow it to cool.
- At 50°C, add 1.5 ul of ETBR. Allow it for cooling.
- Take the clean tray and pour the gel into the tray.
- Place the comb in the poured gel and leave it for solidification.
- After solidification remove the comb from the gel.
- Then add isolated DNA 2-3 ml along with the gel running dye in the well.
- Supply 85 volts and 100 illiampere for 20-30 min.
- Document the gel after running.

General PCR Protocol

1. Prepare the following 50 µl reaction in a 0.5 ml PCR tube on ice:

Component	25 µl reaction	Final Concentration
10X ThermoPol or Standard Taq Reaction Buffer	2.5 µl	1X
10 mM dNTPs	0.5 µl	200 µM
10 µM Forward Primer	0.5 µl	0.2 µM (0.05–1 µM)
10 µM Reverse Primer	0.5 µl	0.2 µM (0.05–1 µM)
Template DNA	variable	<1,000 ng
Taq DNA Polymerase*	0.125 µl	1.25 units/50 µl PCR
Nuclease-free water	to 25 µl	} 30 Cycles

2. Gently mix the reaction and spin down in microcentrifuage.
3. Cycling Conditions for a Routine PCR:

STEP	TEMP	TIME
Initial Denaturation	95°C	30 seconds
Denaturation	95°C	15-30 seconds
Annealing	45-68°C	15-60 seconds
Extension	68°C	1 minute
Final Extension	68°C	5 minutes
Hold	4-10°C	

PCR amplification for Iturin A from Antagonistic *Bacillus*

Forward Primer : 5' GATGCGATCTCCTTGGATGT 3'

Reverse Primer : 5' ATCGTCATGTGCTGCTTGAG 3'

Cycling Conditions for amplification of Iturin A:

STEP	TEMP	TIME	
Initial Denaturation	94°C	3 minutes	
Denaturation	94°C	1 minute	} 40 Cycles
Annealing	60°C	1 minute	
Extension	72°C	1 minute	
Final Extension	72°C	10 minutes	
Hold	4-10°C		

The expected size of the PCR amplified product is 649 bp

PCR amplification for surfactin from Antagonistic *Bacillus*

Forward Primer : 5' ACAGTATGGAGGCATGGTC 3'

Reverse Primer : 5' TTCCGCCACTTTTTCAGTTT 3'

Cycling Conditions for amplification of surfactin:

STEP	TEMP	TIME	
Initial Denaturation	94°C	3 minutes	
Denaturation	94°C	1 minute	40 Cycles
Annealing	57°C	1 minute	
Extension	72°C	1 minute	
Final Extension	72°C	10 minutes	
Hold	4-10°C		

The expected size of the PCR amplified product is 441 bp

15

Agarose Gel Electrophoresis for Detection of Biocontrol Agents

S. Nakkeran and S. Rageshwari

Department of Plant Pathology, Tamil Nadu Agricultural University Coimbatore- 641 003, Tamil Nadu

Agarose gel electrophoresis is used for the fractionation and characterization of nucleic acids. It is a simple and highly effective method for separating, identifying and purifying DNA fragments of various sizes. DNA can be checked for size, intactness, homogeneity and purity by this technique. Agarose is a purified form of agar and forms a gel by hydrogen bonding of the agarose monomers. Increasing the Agarose concentration of a gel reduces the migration speed and enables separation of smaller DNA molecules. Shorter molecules move faster and migrate further than longer ones.

Aim: To study the size of the DNA extracted from the fungal culture.

Principle: Ethidium Bromide added to the gel interacts with the bases of DNA and fluorescence orange when irradiated with UV light.

Materials Required

(a) Tris Acetic acid EDTA (TAE) buffer 5 X (pH 8.3)

(i)	Tris base	: 24.2 g
(ii)	Acetic acid	: 5.71 ml
(iii)	EDTA 0.5M (pH 8.0)	: 10.00 ml
(iv)	Distilled water	: 1000 ml

(Working buffer 1X or 0.5 X TBE)

b) Ethidium bromide solution – 10 mg/ml

0.2 g of Ethidium bromide was dissolved in 20 ml of distilled water and stored at 4°C.

(c) Sample buffer or loading buffer (10 X)

(i)	Sucrose	: 40 %
(ii)	Bromophenol blue	: 0.25 %

[All (W/V in 1 X TAE)]

(d) DNA molecular weight marker (1 kb) (Bangalore Genei)

(e) Agarose – 1.2 %: 1.2g of agarose in 1X TAE buffer

Procedure

- PCR products were resolved in 1.2 % Agarose gel electrophoresis.
- Agarose 1.2 g was taken in a conical flask containing 100 ml of 1X TAE buffer and melted completely into a clear solution using a micro wave oven and the solution was cooled to ear lobe bearable warm and 1 µl of ethidium bromide was added and mixed well.
- The molten agarose was poured inside the preset template sealed with sealing tapes on both the sides of the plate, well forming comb was placed inside the plate containing agarose solution. It was allowed to solidity for 45 minutes.
- Then the comb and sealing tapes were removed carefully and the plate was mounted in an electrophoresis tank (Serva-Germany).
- The tank was filled with 1000 ml of 1X TAE buffer just to immerse the gel up to 1 mm. Ten micro liter of PCR product was mixed with 5 µl of sample buffer and loaded in the well.
- The electrophoresis apparatus was connected to a power pack and the gel electrophoresis was performed at 70 volts.
- The electrophoresis was terminated after one hour. Then the DNA bands were visualized on an Alpha imager (Alpha Innotech Corporation, Germany).

16

Isolation of Secondary Metabolites from Culture Filtrate of *Trichoderma viride* and *Chaetomium globosum*

S. Nakkeran and S. Vinodkumar

Department of Plant Pathology, Tamil Nadu Agricultural University
Coimbatore -641 003, Tamil Nadu

Introduction

Studies on biocontrol agents have revealed the presence of a diverse range of antifungal metabolites derived from amino acid, polyketones and isoprenoid biosynthetic pathways, each of which has distinctive regulatory points and is subjected to different stimuli within organisms. In this experiment, the extraction of secondary metabolites from the culture of *T. viride* and *Chaetomium globosum* will be demonstrated.

Materials

- Solvents, chemicals and glass apparatus
- Laboratory grade reagents such as acetone, hexane, ethyl acetate
- Separating funnel, shaker, conical flasks, sodium sulphate
- Biocontrol agent
- Fungal biocontrol agent *Trichoderma viride*
- Potato Dextrose Broth is used for culturing the *T. viride*

Procedure

- Prepare PDB broth and autoclave it at 15 PSI pressure and 121°C for 20 minutes

- Inoculate 5 mm disk of fungus (*T. viride*) containing spores and mycelia aseptically onto PDB
- Incubate the flask inoculated with *T. viride* at 28°C for 20 days
- During the inoculation, expose the flasks to light for 5-6 h for enhancing sporulation after 15 days on inoculation
- After 20-25 days of inoculation, take culture filtrate in a separating funnel and extract the metabolites with equal volume of ethyl acetate
- Dry the organic solvent using vaccum flask evaporator under reduced pressure
- Spot the dried sample on silica coated aluminium TLC plates
- Use different solvent systems to run TLC in the mobile phase
- Stop the experiment before the mobile phase (solvent) reaches the end point and air dry the plates and observe in UV transilluminator.
- Check the Rf values of sample with the reference samples
- Calculate Rf value using the formula

$$\text{Rf value} = \frac{\text{Distance moved by the solute from the origin}}{\text{Distance moved by the solvent from the origin}}$$

- The spot has to tested against variety of pathogens *in vitro* after purification of the compounds or as such.

Antifungal metabolite extraction from *Chaetomium*

- Grow *Chaetomium* isolates on Potato Dextrose (PD) broth in 250 ml of conical flask containing 100 ml.
- Incubate the flasks at 25°C and 175 rpm for 12 days.
- Filter the culture through two layers of filter paper and extract the filtrates twice with equal volumes of ethyl acetate.
- Evaporate the ethyl acetate on a rotary evaporator and concentrate to dryness
- Dissolve the solid extracts in 1 ml of ethyl acetate
- Load the ethyl acetate extracts on silica gel coated aluminium plates
- Dissolve 100 μg of pure chaetoglobosin A in 1 ml of ethyl acetate and load on silica gel as a standard (reference) sample.

- Elute the sample in mobile solvent phase containing Benzene: ethyl acetate: glacial acetic acid @ 10:1:1 v/v
- Stop the experiment before the mobile phase (solvent) reaches the end point and air dry the plates and observe in UV transilluminator.
- Check the Rf values of sample with the reference samples

PART II
Entomopathogenic Nematodes

17

Introductory Lecture on Entomopathogenic Nematodes

S. Subramanian

Department of Nematology, Tamil Nadu Agricultural University Coimbatore -641 003, Tamil Nadu

Entomopathogenic nematode (EPN) *viz., Heterorhabditis, Neosteinernema* and *Steinernema* have been recovered from many regions of the world. They selectively infect many insects and a few other arthropods but do not adversely affect mammals or plants. They are actually vectors of pathogenic bacteria, cause relatively rapid death of the host (24 – 48 h) and show considerable potential in biological control of insect pests in the integrated pest management systems.

EPNs are similar to insect parasitoids in that the immature form develops at the expense of one host individual; they are ecologically similar to both parasitoids and arthropod predators because the host individually killed. However, these nematodes differ from predators and parasitoids in two ways: (1) the nematode is mutualistically associated with a bacterial pathogen which actually kills the host, and (2) the pathogenicity of the bacterium has a great influence on the efficacy of the system (Ehler, 1990). The purpose of this chapter is to introduce EPN as biocontrol agents in this refresher course and to provide an overview of some current issues in biological control of insects with particular emphasis on their relevance to the use of EPNs.

Biological control is defined as the action of natural enemies which maintains a host population at levels lower than would occur in the absence of the enemies. Natural enemies will include parasitoids, predators, nematodes, microbial pathogens and their gene products. Entomologists generally divide biological control into two categories : (1) natural biological control which is effected by native or coevolved natural enemies in the native home (origin) of a given insect and (2) applied biological control which due to human intervention. Applied biological control is further divided into classical biological control (introduction of exotic natural enemies) and augmentative biological control (enhancement of natural enemies already in place).

Natural Biological Control

It is important to study and document naturally occurring biological control. The knowledge gained from such studies can be relevant to both applied biological control and pest management by EPNs.

The role of EPNs in natural biological control of insect pests needs thorough investigation wherever possible. The native home of a given nematode should be determined to carry out ecological investigations in the habitat in which the nematode species actually evolved. At the generic level, the centres of origin for *Steinernema* and *Heterorhabditis* should be determined because evolutionary theory suggests the presence of closely related species in the center of origin of a given genus. Such species could have considerable potential in applied biological control of insects.

Classical Biological Control

The importation of exotic natural enemies for biological control of both native and exotic insect pests has been followed for more than 100 years. The practice of classical biological control is well developed and consists of the following sequence of procedures: foreign exploration (in native home of the pest), quarantine of imported organism, mass production of the candidate, field colonization and evaluation. Some protocols for importation of insect – parasitic /pathogenic nematode were provided by Nickle *et al.* (1988). Poinar (1986) stated that a few nematodes have been employed in classical biological control and with the current interest in EPN species, it is likely that the number of introductions will increase considerably.

Augmentative Biological Control

Human intervention planned to enhance or augment the effectiveness of natural enemies already in place is applicable to both native and exotic species. These augmentative procedures involve manipulation of environment or natural enemy or both.

Coppel (1986) recognized nine areas for environmental management or manipulation. They are (i) land use; (ii) habitat provision; (iii) food provision (iv) following tillage methods for annual crops; (v) modified release strategies (vi) host or prey provision (vii) reducing enemies of natural enemies (viii) improved pesticide utilization and (ix) use of semio chemical and behavioural strategies.

The direct manipulation of a natural enemy consists of mass production and field release of individuals of a given spaces. Entomologists recognize two kinds of releases *viz.*, inoculative release (season long pest suppression) and inundative release (immediate pest suppression).

For augmentative release of EPN nematologists are encouraged to consider two areas *viz.,* 1. Inundative release of nematodes and 2. Genetic improvement of natural enemies as suggested by Gaugler (1987, 1989) for pesticide resistance, improved tolerance to the soil environment, increased retentiom of symbiotic *Xenorhabdus* / *Photorhabdus* bacteria.

Environmental Impact

In the recent years, the subject of environmental impact of introduced species of natural enemies have emerged as one of the major concerns. From a broad ecological perspective, introduced natural enemies such as predators and parasites of insect pests and those phytophagous insects imported for weed control can be expected to have an environmental impact. The same assumption should hold good for entomopathogenic nematodes also.

a. Ecological Perspective

Environmental impact can be defined as any effect on a non target organism which results from the intentional introduction of a natural enemy. Thus, an introduction resulting in permanent establishment of a biological control agent will probably have an environmental impact; however, the impact may not necessarily be an adverse one. Impacts can be classified according to a number of factors such as duration, predictability, outcome, magnitude, interaction and timing.

b. Entomopathogenic Nematodes

We cannot say that releases of entomopathogenic nematodes will have no environmental impact. The environmental impact of a biocontrol agent is a function of (i) the attributes of the introduced species (ii) the nature of the target zone and (iii) the introduction strategy employed. Entomopathogenic nematodes may pose little environmental risk because of their minute / microscopic size, although considerably more evidence may be required before a generalization can be made.

Conclusions

Modern insect control is shifting away from dependance on synthetic organic insecticides in favour of a more integrated approach to pest management. Biological control is a major tactic in integrated pest management (IPM).

Entomopathogenic nematodes are the latest addition to the natural – enemy pool and should further enhance the ability to fully integrate the various management measures for target pests where individual control tactics alone are inadequate. Biocontrol scientists must seek and work for optimal allocation of effort towards biocontrol and IPM so that ecologically sound pest control can be maximized.

References

Ehler, L.E 1990. Planned introductions in biological control in Assessing. Ecological risks of Biotechnology. Ginzburg, I.R., Ed. Butterworths, Boston.

Poinar, G.O., Jr., 1986. Entomophagous nematodes, in Biological Plant and Health Protection, franz, J. M. Ed., G. Fischer Verlag, Stuttgart.

Nickle, E.R., Drea, J. J., and Coulson, J. R., 1988. Guidelines for introducing beneficial insect – parasitic nematodes into the United States, *Ann. App. Nematol.,* 2: 50.

Coppel, H. C., 1986. Environmental management for furthering entomophagous arthropods, in Biological Plant and health Protection, Franz, J. M. Ed., G. Fischer Verlag, Stuttgart, 57.

Gaugler, R., 1987. Entomogenous nematodes and their prospects for genetic improvement in Biotechnological Advances in Invertebrate Pathology and Cell Culture, Maramorosch, K., Ed., Academic Press, New York, 457.

Gaugler, R., Campbell, J. F., and McGuire, T.R., 1989. Selection for host-finding in *Steinernema feltiae, J. Invertebr. Pathol.,* 54: 363.

18

Mass Production of *Corcyra cephalonia* Stainton and Greater wax moth, *Galleria mellonella*

S. Sridharan, P.A. Saravanan and T. Manoharan

Department of Agricultural Entomology, Tamil Nadu Agricultural University Coimbatore- 64103, Tamil Nadu

Introduction

Corcyra cephalonica commonly called as rice meal moth or rice moth is a pest of stored foods, *viz.,* cereals, cereal products, oilseeds, pulses, dried fruits, nuts and spices. Many of the natural enemies mass-bred in the laboratory for use in field against crop pests are dependent on either egg or larval stages of *Corcyra* due to the simple reason that it is easier and cheaper to produce natural enemies on different stages of *Corcyra* than on their original hosts.

Morphology and Biology of *Corcyra*

The eggs are oval and measure 0.5 x 0.3 mm. The white surface is sculptured and has a short nipple-like structure at one end. The larvae are generally creamish – white except for the head capsule and the prothoracic tergite, which are brown. There are well-developed prolegs on abdominal segments 3-6 and 10. A fully matured larva measures 15 mm. The last-instar larva spins a closely woven, very tough, double-layered cocoon in which it develops into a dark-brown pupa. The anterior portion of the cocoon has a line of weakness through which the adult emerges. The adults are small. The hind-wings are pale-buff, and the fore-wings are mid-brown or greyish-brown with thin vague lines of darker brown colour along the wing veins. The males are smaller than the females.

Sexual activity usually begins shortly after adult emergence. There is a pre-oviposition period of about 2 days. Egg-laying mainly occurs during the night.

The greatest numbers are laid on the second and third days after emergence, although oviposition may continue throughout life. Eggs take about 2-3 days to hatch. Optimum conditions for larval development of *C. cephalonica* are 30 – 32.5 °C and 70 per cent RH, at which, the period from egg hatch to adult emergence is only 26-27 days. There is considerable variation in the number of larval instars; however, males generally have 7 and females have 8. The last-instar larvae pupate within the food. The adults emerge through the anterior end of the cocoon, where there is a line of weakness. The sex ratio is 1:1. The adult moth is nocturnal and is most active at nightfall.

Mass Production of *Corcyra* in the Laboratory

Materials required

Absorbent cotton	Storage racks
Blotting paper	Streptomycin sulphate
Broken cumbu grain	Rubber band
Camel hair brush	Measuring cylinder
Enamel Tray	Oven
Honey	Home milling machine
Khada cloth	Sieves
Mosquito net	Formaldehyde 40%
Moth aspirator (collector)	Filter paper
Oviposition drums	Moth scale egg separator
Plastic basin	Face masks
Shoe brush	Storing drums
Soap	Ground nut kernel
Specimen tube	Sulphur (WP)
Yeast	Coarse weighing balance

Procedure

Preparation of Rearing Basins

The basins (16" dia) used for *Corcyra* multiplication are thoroughly cleaned with 0.5% detergent wash and rinsing in tap water followed by wiping with dry, clean – used towel and shade drying. Whenever the trays are emptied after a cycle of rearing, they have to be cleaned preferably to 2 per cent formaldehyde and returned to storage until further use. On reuse, the cleaning steps are repeated.

Preparation of Bajra Medium for *Corcyra*

a. The required quantities of bajra grains are coarsely milled and broken into 2-3 pieces in a milling machine. The broken grains are heat sterilized at 100°C for 1 hour to eliminate the residual population of stored product insects *viz., Rhizopertha dominica, Sitotroga cerealella, Tribolium castaneum* and fungal contaminants. Upon sterilization the grains are cooled under fan in a clean area. The grains are then transferred to plastic basins @ 2.5 kg/basin.

b. Groundnut kernel in required quantity is broken using a pounding machine or a mechanical blender (domestic mixer). Then 100 g of the broken kernel is transferred to each basin and the contents are hand mixed thoroughly.

c. Dry yeast (Bakers) and wettable sulfur is added @ 5g/ basin and the contents are mixed thoroughly. A spray of 10 ml of 0.01-0.05% streptomycin sulfate and mixing of the contents follows this. This medium is used for rearing *Corcyra* larvae

d. The number of basins required for egg infestation is calculated and the medium is prepared accordingly.

Preparation of *Corcyra* Eggs

The primary source of *Corcyra* eggs is reputed laboratories, commercial producers for bulk preparation. If it is intended to begin the production with nucleus colony, the adult moths can be collected from warehouses where the food materials are stored.

a. The eggs used for building up the colony of *Corcyra* have to be free from contaminants like the moth scales and broken limbs and not exposed to UV light.

b. The collections of overnight laid eggs are measured volumetrically to ascertain the number of trays that can be infested with eggs. A cc of eggs is known to contain approximately 16000 – 18000 eggs.

Infestation of Medium with Eggs

The overall production scheme involves initial infestation of the cumbu medium with *Corcyra* eggs in desired quantities. This is accomplished by sprinkling the freely flowing eggs on the surface of the medium in individual basins. Per basin 0.5 cc eggs of *Corcyra* is infested. The basins are then covered with clean *khada* cloth and held tightly with rubber fasteners. The basins are carefully transferred to the racks. At a time 84 such basins are stacked in the rack designed at TNAU.

Handling the Trays During Larval Development

The larvae that hatch out in 3-4 days begin to feed the fortified Bajra medium. At this stage, light webbings are noticed on the surface. As the larvae grow up they move down. During this period the larvae are allowed to grow undisturbed in the trays.

Handling of Adults

The adults begin to emerge in 28-30 days after infestation of the eggs. The adults can be seen on the inner side of the *khada* cloth. They are either aspirated with mechanical moth collector or collected with specimen tubes. The whole operation is carried out in a tent of mosquito net. This prevents the large-scale escape of the moths, which if uncontrolled can migrate to the storage area and spoil the grains stored by laying eggs. Workers involved in the collection of moths should wear face moths continuously to avoid inhalation of scales. The moths collected are transferred to the oviposition drum @ 1000 pairs per drum at a time. The oviposition drums of size 30 x 20 cm are made of galvanized iron. The drums rest on tripod frames with legs of height 5cm. The bottoms of the drums are provided with wire meshes that enable collection of eggs. The walls of the drums have two vents (ventilation holes) opposed to each other. The vents are again covered with wire mesh. The lids of the drums have handles besides slots for introducing the moths and adult feed. The oviposition drums filled in a day are maintained for four to seven successive days for egg collection after which are emptied and cleaned for next cycle of use.

The adults are provided feed containing honey solution. The adult feed is prepared by mixing 50 ml honey with 50 ml water and 5 capsules of vitamin E (Evion). The feed is stored in refrigerator and used as and when required. Piece of cotton wool tied with a thread is soaked in the solution and inserted into the drum through the slot at the top. From a basin, moths can be collected upto 90 days after which the number of moths emerging dwindles down and keeping the basins is not economical for the producer.

Handling of Eggs

The moths lay the eggs in large numbers loosely. The scales and broken limbs are also found in larger quantities along with the eggs. They cause potential hazard to the workers after years of working in *Corcyra* laboratory. To minimize the risk of scales freely floating in the air, the oviposition drums are placed on sheets of filter paper in enamel trays which trap effectively the scales. Sets of several oviposition drums are kept in ventilated place near an exhaust fan to enable the workers comfort. Daily morning the oviposition drums are lifted up and the wire-mesh bottoms are cleaned gently with a shoe brush so that the eggs and remnants of scales and limbs settled on the mesh are collected along with those on the filter paper. The collections are cleaned by gently rolling the eggs on filter paper to another container. Then they are passed to sieves in series and finally clean eggs are collected. The eggs are quantified in measuring cylinders and used for building up the stocks and natural enemy production.

About 100 pairs of adults produce 1.5 cc of eggs in 4 days laying period inside the oviposition drums. From each basin an average of 2500 moths are collected. Hence from each basin 18.00 – 20.00 cc of eggs can be obtained in 90 days.

Problems Encountered in Mass Production of *Corcyra cephalonica*

Introduction

The mass production of *Corcyra* involves materials and production management and assurance of quality. The rearing area in the *Corcyra* laboratory has to be maintained in clean condition and free from dust always. The larvae have to be confined within the basin and escape of larvae and adults compound the problems to the producer. Therefore, the producers must have prior knowledge on the problems encountered during culture and methods to manage.

Problems Encountered

The bajra medium used for *Corcyra* rearing is attractive to other insects and they compete with *Corcyra* for the food and the yield of adults from individual basins gets reduced. In addition, the mite, *Pyemotus ventricosus* parasitic on the *Corcyra* multiply in large numbers and debilitate the larvae. Hence, there is potential danger of losing the culture.

Tribolium castaneum (Herbst) (Red flour beetle)

Family: Tenebrionidae; Order: Coleoptera

T.castaneum is cosmopolitan in occurrence. The eggs of *T.castaneum* are approximately 0.5 mm long, cylindrical and white. They are covered with a sticky secretion, which causes them to become covered in flour and stick to containers. The larvae are yellowish-white, cylindrical and covered with fine hairs. The head is pale-brown and the last segment of the abdomen has two upturned dark, pointed structures. The pupa is naked (without a cocoon), yellowish-white, becoming brown later. The dorsum is hairy and the tip of the abdomen has two spine – like processes. The adult is 2.3 – 4.4 mm long, rather flat, oblong and chestnut – brown (reddish-brown). The head and upper part of the thorax are covered with minute punctures and the wing covers are ridged lengthwise. Adult females of *T. castaneum* lay up to 450 eggs in stored products. The incubation period of the eggs is between 5 and 12 days. The grubs and adult stages of *T. castaneum* infest the host. The fully grown grubs are 6 mm long and pupate in 27-29 days. Pupation occurs in the host medium. The pupa of *T. castaneum* is naked (without a cocoon). Adults emerge from the pupa within 7 days. The adults may live for as long as 18 months, depending on weather conditions. Adults fly in large numbers in the late afternoon.

Detection of Infestation

At low levels of *T. castaneum* infestation, the medium may not show any sign of holes or tunnels. At high levels of infestation, holes and tunnels can be seen. Adults are present on the walls of the basins; larvae inside the grains; and eggs are covered with flour or dust and are found sticking to basins. In the event of outbreak, the medium becomes mouldy and emit a pungent odour. The adults can be seen during the afternoon hours congregating on the walls of *Corcyra* laboratories. More than the presence in the medium, the beetles are major sources of irritation to the workers. They get into the garments and cause inconvenience.

Losses Caused

Tribolium grubs effectively compete with the *Corcyra* larvae for feeding and displace them. In this process, there is potential loss in growth of the *Corcyra* larvae. When the infestation reaches alarming proportions there is a corresponding decrease in the moth yield per basin.

Management

Preventative control of *Tribolium* is better. This is achieved by selection uninfested stock bajra grains from the market, good house keeping, sanitation in the rearing environment, avoidance of spills of grains in the rearing and storage areas, heat sterilization of the grains during media preparation etc. In addition, the area where the insects are housed should not have cracks and crevices as they act as good hiding places for the adult insects.

Timely management has to be taken when infestations are noticed. Though fumigation methods are available they can be practiced if only the basin with entire media and insects are to be disposed off. More over it requires care and supervisory control has to be followed. The detection of insects in bulk storage is difficult and conventional methods are insensitive to low population densities. All the techniques developed are so far aimed at management of the adults only. Trapping techniques appear to offer a more effective approach to pest detection in bulk media.

1. Paper boards of size 10x10 cm are placed on the surface of the media a week after infestation of *Corcyra* eggs in the basins. The paper strips are observed daily. The adults clinging to the paper strips are collected manually and killed in water containing 0.5 per cent teepol (surfactant).
2. The adults are attracted to near UV rays. After dusk 4 watts UV lamps are setup near the racks where the basins are kept. A basin containing water and 0.5 per cent teepol is placed below. The *Tribolium* adults are attracted towards the light and trapped in water. The traps are to be operated when all the activities are completed and no worker is allowed to enter the lab as the rays are harmful to workers.
3. Flour traps containing 250 g whole wheat flour and 5 g dry yeast is spread in basins and placed on the floor below the racks @ 1-2/rack. These flour traps are good oviposition attractants for the beetle. Alternate days the flour in the basins is sieved using a fine mesh sieve and the eggs and adults in the flour are collected and destroyed in 0.5% teepol.
4. The adults found on the walls are removed with vacuum cleaner regularly.

***Bracon hebetor* Say**

Family: Braconidae; Order: Hymenoptera

They occur sporadically in the rearing laboratory. They are larval parasitoids and pupate in groups outside the body of the host insect. The adults can be seen hovering in the lab or seen settled on the top of *khada* cloth fastened onto the basin. These parasitoids complete developmental period in Corcyra larvae in 7-10 days. The oviposition lasts for 13 days. During its life a *Bracon* female can lay 100-150 eggs and a single adult can parasitize upto 32 *Corcyra* larvae.

Detection

The adults are small and have long antennae. The larvae that are parasitized become paralyzed and undergo discoloration. When the grubs emerge out for pupation, the host larvae die, turn black and become scaly. From the cocoon formed, next generation of adults emerge and continue the damage caused in previous generation.

Losses Caused

The *Bracon hebetor* is a contaminant in *Corcyra* culture. It can cause loss to the extent of 50 per cent in production. The loss will be all the more high if *Tribolium* infestation occurs together with *Bracon*.

Management

1. Total management of the population of *Bracon* is needed to eliminate the infestation. The basins that are used for rearing *Corcyra* should be disposed after 90 days. After this period the food inside the basin is either exhausted or becomes unsuitable for the remaining larva inside. At this stage, migration of the larvae starts. This is made easier when the *khada* cloth used for covering either has holes or is too saggy and touches the surface of the medium. The migrant larva and those found inside or outside the *khada* cloth are the primary sources for parasitization. In a short period the attack becomes alarming. Therefore, enough care has to be taken to maintain the basins. The migrant larva if any found on the walls have to be removed and killed. The basins should be examined periodically and they should not be kept open except during observation and adult handling. During observation discoloured larvae, white cocoons if found are collected and immersed in 0.5% formaldehyde to kill the parasitoids.

2. The adults of *Bracon* are highly phototropic and attracted to incandescent light. In the *Corcyra* laboratory, during night hours a table lamp with 60W – tungsten bulb is set up over a basin with water. The adults are attracted easily and killed. The bulbs can be operated regularly in the laboratory.

3. As a routine practice, all the materials in the *Corcyra* laboratory are given a spray of 0.1 per cent malathion. Care should be taken to apply the pesticide even on to the furniture and basins. When the problems of *Bracon* turn acute disinfection of all the basins are done one by one in separate room. The rearing area is completely sprayed with malathion and the doors, windows and exits are closed for thee days. The laboratory is cleaned thoroughly before transport of the rearing basins and other materials back to the rearing area.

The greater wax moth or honeycomb moth (*Galleria mellonella*) is a moth of the family Pyralidae. It is found in most of the world, including Europe and adjacent Eurasia, its presumed native range, and as an introduced species on other continents, including North America and Australia. The greater wax moth larva, attacks the honey bee comb and bee products in the colony. Less often, they are found in bumblebee and wasp nests, or feeding on dried figs. The waxworms of the greater wax moth have been shown to be an excellent model organism for *in vivo* toxicology and pathogenicity testing, replacing the use of small mammals in such experiments. The larvae are also well-suited models for studying the innate immune system.

Biology

The female moth can lay 50 to 250 eggs during its life time. The egg period varied from 3-6 days whereas the 1st, 2nd, 3rd, 4th and 5th instar larval period complete in 3-6, 4-7, 4-7, 5-10 and 7-10 days respectively. The pupal period is 10-14 days and the adult longevity of the male and female is 0.5 and 2-9 days respectively.

General Facilities

Two rooms, 3.05 by 2.44 by 2.55 m, can be used for larval rearing and oviposition and a work area, 3.05 by 4.57 m, outside the rooms is necessary for producing eggs *of Galleria* at the rate of about 1 million eggs per day. Each room should have lights, a power source, and equipment for maintaining 65±5 percent relative humidity (RH) and 30±2 °C. For larval rearing, there are three tiers of shelves, each 46 cm deep and 38 cm apart starting 38 cm from the floor on two sides of the room. Water, power, a floor drain, and at least 3.34 m^2 of bench space are needed in the work area. A refrigerated area for storing diet ingredients is also desirable. For oviposition, boxes with shelving to hold pans of larvae and pupae are required along with the special oviposition cages containing plates from which egg collection can be made at planned intervals.

Larval Rearing

Larvae can be reared in any sort of pan or jar. It is most efficient to use 4.4-liter glass jars containing 2.2 liters of artificial diet for the first 2 weeks of larval development because they require little space. Then, larvae transferred to round, galvanized iron pans (11.5 liters) with more diet. Larvae require about 2 weeks to complete development, at 30 °C, after transfer to the larger pans. It constructs the cocoons in the diet around the edges of the pans.

The following diet is recommended for mass rearing *Galleria:* Wheat flour 350 g, corn flour 200 g, milk powder 130 g, backing yeast powder 70 g, honey 100 ml, and glycerin 150 ml.

Liquid ingredients of the diet are mixed and added to the remaining dry ingredients that have been previously mixed then added to provide even distribution of the cornmeal flour, and wheat flour. The diet may be mixed in any large container or barrel, but a 56.6 liter capacity cement mixer greatly reduces the labor. The number of larvae necessary to infest the diet can vary widely without grossly affecting egg production. A decrease in larval density results in an increase in size of adult females and in higher egg production per female. The optimum is about 1,100 eggs/kg (= 37.6 mg).

References

Kennedy J.S. and Zadda Kavitha 2007. Commercial production of biocontrol agents Department of Agricultural Entomology Directorate of Centre for Plant Protection Studies Tamil Nadu Agricultural University, Coimbatore P.153

19

Taxonomy and Identification of *Steinernema* and *Heterorhabditis*

M. Sivakumar and S. Subramanian

Department of Nematology, Tamil Nadu Agricultural University
Coimbatore -641 003, Tamil Nadu

Although the Rhabditids have been studied from last one century, the field level use of Entomopathogenic nematodes the members of Rhabditids for the management of insect pest was followed only after Dutky who practically used EPNs in 1937 on Apple Codling moth.

The taxonomy of Entomopathogenic nematodes of the order Rhabditidae is discussed in this lecture.

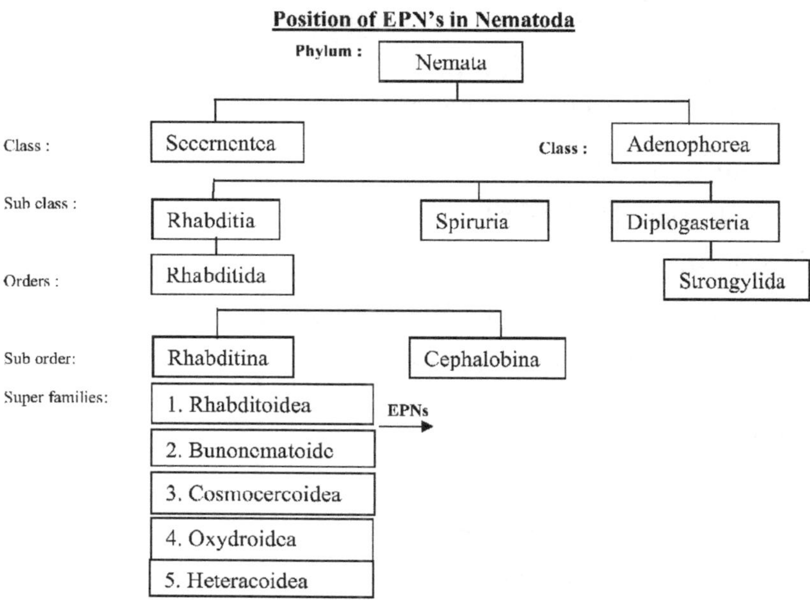

Sub class Rhabditia

- Rhabditoid oesophagus; (Crescent valve at basal bulb)
- Immature females – oesophagus 3 parted *viz.,* Corpus, Isthmus, and basal bulb
- Adult parasitic – Oesophagus clavate or cylindrical

Order : Rhabditida

Labia : 6,3,2 or none

Stoma : Tubular with 5 rhabdions

Oesophagus : 3 parted with butterfly valve in basal bulb.

Excretory system: Excretory tube two, cuticular lined, paired lateral collecting tube runs posteriorly.

Reproductive system: One or two ovaries, vulva posterior if one ovary is present

Rhabditida has 2 sub orders *viz.,* **Rhabditina and Cephalobina**

Sub order Rhabditina

- Labia distinct and have papilloid cephalic sensilla and pore like amphids
- Stoma cylindrical and devoid of rhabdions 2 or more times as long as wide
- Oesophagus 3 parted with valved muscular basal bulb
- Females with one or two ovaries, males with paired spicules, and gubernaculum, caudal alae present.

Super family : Rhabditoidea

- Well developed cylindrical stoma
- Labia 2-6
- Oesophagus with roller type valve in basal bulb
- Caudal alae with 5-9 papilloid rays

Taxonomy of EPNs

Phylum	:	Nematoda
Class	:	Secernentiea
Order	:	Rhabditida

Sub order	:	Rhabditina
Super family	:	Rhabditoidea
Family	:	Steinernematidas
Type genus	:	*Steinerma* Syn : *Neaoplectana* Steiner, 1927
		Type species : *Neaoplatana glaseri*
Other species	:	*Steinernema glaseri*
		S. bibionis
		S. feltiae
		S. intermedia
		S. rara
		S. affinis
		S. anomal
Family	:	Heterorhabditidae
Type genus	:	*Heterorhabditis*
Type species	:	*H. bacteriophora*
Other species	:	*H. heliothidis*
		H. megidis

Steinernema sp.

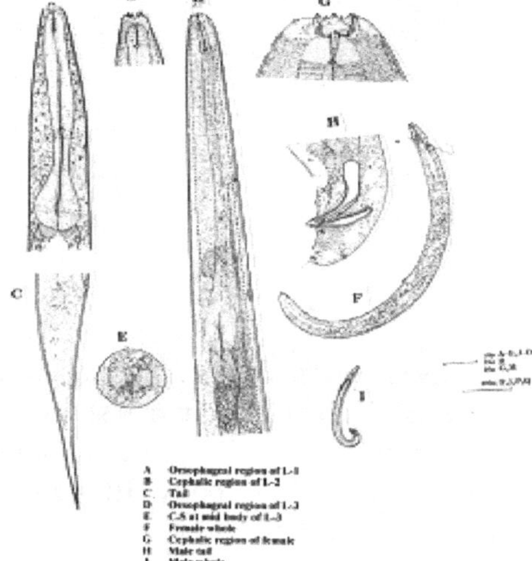

Heterorhabditis sp.

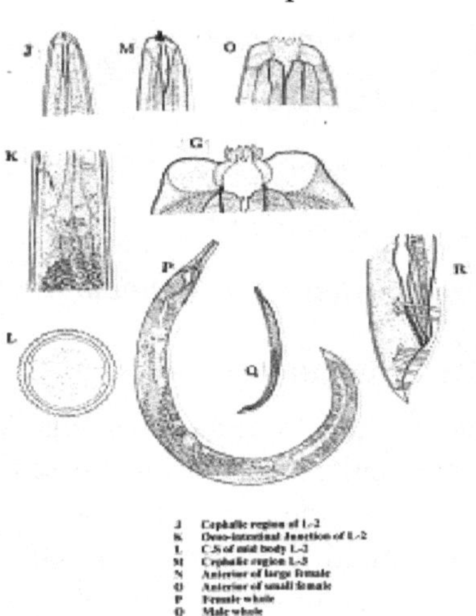

Morphology of *Steinernema* and *Heterorhabditis*

1. Steinernema and Neosteinernema

EPN – III stage infective (dauer) often enveloped in II cuticle (moulted), III stage contains symbiotic bacterium in alimentary tract (*Xenorhabdus*).

Adults : Amphimictic Head rounded, not offset. Stoma partially collapsed nerve ring surrounding isthmus. Spherical basal bulb with a reduced valve ventral excretory pore distinct. Lateral field and phasmids inconspicuous.

Females : Amphidelphic with reflexed ovaries functional vulva, size variable. Excretory duct with large glands.

Male : Testis single reflexed spicules paired. Gubernaculum present several pairs of genital papillae present, pronounced ventral preanal papilla, Bursa absent.

Juvenile: II stage has closed stoma. III stage infective dauer juvenile also with closed stoma. Tail sharp and tapering field with 9 lines. Excretory pore above the nerve ring.

Important species

1. *S. abbasi* (= *S. thermophilum)*
2. *S. anamoli*
3. *S. carpocapsae*
4. *S. feltiae*
5. *S. glaseri*
6. *S. intermedium*
7. *S. kushidai*
8. *S. riobravi*
9. *S. siamkayai*
10. *Neosteinerma longicurricauda*

2. Heterorhabditis

Infective stage with closed stoma. Morphology similar to Steinernematidae. III stage juvenile infective and posses a characteristic labial tooth above stoma on the dorsal side. An indepentant ventral screrotized labial plate present. Longitudinal lines cover the entire circumference of the body. III stage – Alimetary tract with symbiotic bacterium. *Photorhabdus*. Males have long spicules. Bursa supported by ribes. Genital papille absent. Males produced only in second generation. Males are considerably smaller than females.

Important species

H. bacteriophora, H. indica and *H. megidis*

References

Gaugler, R. and Kaya, H.K 1990. Entomopathogenic nematodes in Biological control. CRC Press.

Gaugler, R 2002. Entomopathogenic nematodes, CABI Publishers, UK.

20

Microscopic Examination and Identification of *Steinernema* and *Heterorhabditis*

K. Poornima

Department of Nematology, Tamil Nadu Agricultural University
Coimbatore -641 003, Tamil Nadu

Basically we need to mount and prepare the slide for observation of various stages of EPN as detailed below in Table.1

Table 1. Stages of EPN and suitable slides

S.N	Genus stage	Slide preferred
1.	3rd stageDauer infective	Glass slide (glycerine mount)
2.	Adult male	Glass slide (glycerine mount)
3.	Adult female	Glass slide (Wax mount)

For identification of EPNs many morphological characters are compared for differentiation and furnished below in Table 2.

Table 2. Morphological differences between *Steinernema* and *Heterorhabditis*

S.No	Characters	*Steinernema*	*Heterorhabditis*
1.	II stage cuticle retention by III stage infective juveniles	Easily lost	Good retention
2.	III stage cuticle visible around infective juvenile	When present, loosely fitted around infective juvenile	Tightly fitted around infective juvenile, difficult to see
3.	Infective juvenile develops into	Adult male or female	Hermaphrodite
4.	Location of bacterial symbiont	Ventricular portion of intestine	Throughout intestine

5.	Luminescence by bacterial symbiont	No	Yes
6.	Cadever colour	Ocher, black	Purple, Red, green, orange
7.	Reproduction	Sexual (Amphimictic)	Hermaphrodite in 1st generation and sexual in 2nd generation
8.	Bursa in male	Absent	Present
9.	Excretory pore position in infective juvenile	Above nerve ring	Below nerve ring

Certain precautions are to be taken while making slides for the observation EPN which are listed below.

1. Use anhydrous glycerine and minimum quantity should be kept in slides.
2. Use glass rods and not glass fibre (preferably glass bangle can be used for making glass rod). Keep glass rods in desiccator submerged in glycerine.
3. For ceiling the slides, use quick fix or anabond instead of cutex or nail polish.

Reference

Subramanian, S., Saravanapriya, B., Abirami, R. and Priya, P. 2005. A practical guide for entamopathogenic nematodes. Dept. of Nematology, TNAU, Coimbatore-3 18 pp.

21

Mass Multiplication of Entomopathogenic Nematodes by *in vivo* and *in vitro* Methods

N. Swarnakumari

Department of Nematology, Tamil Nadu Agricultural University
Coimbatore -641 003, Tamil Nadu

I. *In vivo* culturing technique

In vivo culturing means culturing of entomopathogenic nematodes (EPN) in the host. Generally rice meal moth, *Corcyra cephalonica* and greater wax moth, *Galleria mellonella* are used as insect hosts for rearing EPNs.Insect hosts such as *Spodoptera*, *Helicoverpa* and *Bombyx mori* are also used for culturing *Steinernema* and *Heterorhabditis* but these insects need plant host especially leaves for culturing. *Corcyra* can be grown in broken cumbu grains and *Galleria* in honey comb or artificial diet.

Mass culturing

- Collect fourth instar larva.
- Inoculate infective juveniles (20 IJ/larva) of *Steinernema* or *Heterorhabditis*.
- Incubate for 2 – 3 days for infection.
- Infected larva becomes brick red colour if infected with *Heterorhabditis*.
- Place the infected larvae on White's trap.

White's trap method for isolation of IJ of EPN

- Take a large Petri dish (20 cm).
- Take watch glass and place it in the bottom of Petri plate.

- Cut a Whatman No.1 filter paper into 2 to 3 pieces according to the size of watch glass.
- Place filter paper on the watch glass.
- Add sterile water in the Petridish.
- The filter paper should touch the water.
- Keep the infected larvae on the filter paper.
- Close the bottom of Petriplate with same size as lid and seal it with klinfilm, parafilm or cello tape.
- IJ will be collected in the water present in the Petri plate.

Special requirement for *Steinernema glaseri*

Modified White's trap

- Instead of watchglass, a small Petri plate (5cm) is filled with Plaster of Paris.
- Place the plate with plaster of Paris in the bottom of Petri plate. Wet the surface of plaster of Paris sparingly.
- Place the EPN infected larva on the Plaster of Paris.
- Add sterile water in to the Petridish.
- Infective juveniles will be collected in the water after crawling on Plaster of Paris.

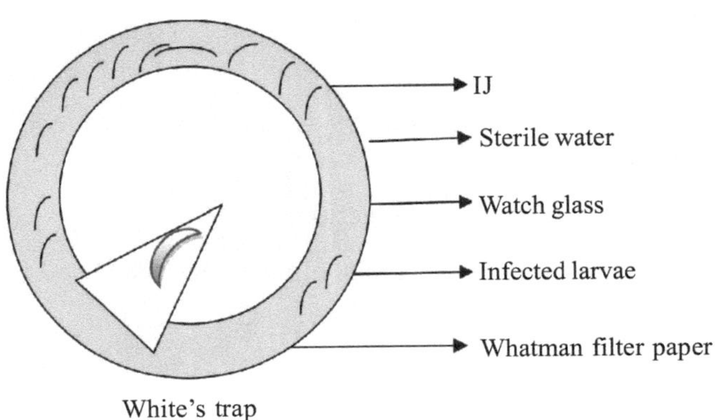

White's trap

II. *In vitro* culturing method

Multiplication of nematodes using artificial media is called *in vitro* culturing. A standard medium developed by Wouts (1981) is commonly used for *in vitro* method.

- Collect anEPN infected insect larva.
- Surface sterilize the infected larva with sterile water.
- Streak haemolymp of insect larva in Nutrient Agar plates.

  0.8 % Nutrient broth

  1.2 % Agar
- Incubate plates at 25° C for 3 days.
- Sub culture repeatedly until a pure culture is obtained.
- Prepare fortified Nutrient agar medium by adding 1% corn oil for fast growth of the bacterium.
- Remove a small piece of agar with bacteria.
- Inoculate sterile IJ of EPN (20 IJ / g) in the cavity of agar plate.
- These IJ move towards the bacterial culture and developed into adult female within 3-5 days.
- Recover nematodes after two weeks of inoculation (*S. glaseri* may be recovered within 10 days of inoculation).

Mass culturing

1. Wouts Media

- Prepare Wouts solid media as per the following ingredients.

Nutrient broth	: 44.0 g
Yeast extract	: 16 g
Soy flour	: 72 g
Corn oil	: 52 ml
Distilled water	: 270 ml

- Dissolve nutrient broth and yeast extract in water.
- Mix corn oil and soy flour and cook.

- Squeeze the media 1g foam cubes ($1cm^3$) to 100 g medium.
- Transfer the foam cubes to conical flasks.
- Sterilize the media impregnated foam in autoclave at 121°C for 15 minutes at 1.5lbsPsi.
- Prepare bacterial cultures in peptone water.
- Inoculate 1 ml of bacterial culture / foam cube.
- Incubate at 25°C for 3 days.
- Inoculate 100 IJ of EPN per flask.
- Incubate the flasks in dark at 25°C for 10 – 15days.
- Transfer the contents in flask to a sieve (40 mesh).
- Wash the foam with sterile water and collect in another container.
- Transfer the suspension in conical flask and allow settling.
- Decant the supernatant and repeat for three times.
- Store cleaned suspension in 250ml canted tissue culture flask @ 20°C in a BOD.

2. Other media

EPN can be cultured in oil cakes and animal origin media also. The following table shows different combinations of media with animal origin. Among which dog biscuit is easy to prepare.

S.No	Animal origin / cake	Water	Foam
1.	Chicken / goat homogenate – 4g	25ml	1g
2.	Dog biscuit – 4 g	25ml	1g
3.	Groundnut oil cake – 4g	25ml	1g
4.	Gingelly oil cake – 4g	25ml	1g
5.	Castor oil cake – 4g	25ml	1g

- Prepare media according to the table given.
- Inoculate and culture EPN as per method describe in procedure 1.

References

Wouts, W.M. 1981. Mass production of the entomopathogenic nematode, *Hetero Rhabdirtisheliothidis* (Nematoda: Heterorhabditidae) on artificial media. *Journal of Nematology,* 13 (4): 467-469.

David I. Shapiro-IlanRichou Han, and Claudia Dolinksi. 2012. Entomopathogenic Nematode Production and Application Technology. *Journal of Nematology*, 44(2): 206-217.

Hanan A. Elsadawy. 2011. Mass production of *Steinernema* spp. on *In vitro* Developed solid medium. *World Applied Sciences Journal,* 14 (6) : 803 – 813.

22

Basic Laboratory Methods Associated with Entomopathogenic Nematodes

P. Vetrivelkalai

Department of Fruit Crops, Tamil Nadu Agricultural University
Coimbatore - 641 003, Tamil Nadu

Introduction

Entomopathogenic nematodes (EPNs) *viz*., *Heterorhabditis, Neosteinernema* and *Steinernema* have been recovered from many regions of the world and have been used commercially as biocontrol agents of insect pests. They are associated with mutualistic bacteria in the genus *Xenorhabdus* for Steinernematidae and *Photorhabdus* for Heterorhabditidae. The infective juvenile nematodes enter their host insect through natural openings (mouth, anus or spiracles), they release bacteria into the insect and cause rapid death of insect host (24-48 h). EPNs were used against many insect pests *viz.,* rice leaf folder, *Cnaphalocrocis medinalis* (Srinivas and Prasad, 1991); tobacco cutworm, *Spodoptera litura* (Rajkumar *et al.,* 2003); brinjal fruit borer, *Leucinodes orbonalis* (Hussaini *et al.,* 2002) diamond back moth, *Plutella xylostella* (Singh and Shinde, 2002); *Diaprepes abbreviates* (Jenkins *et al*., 2007); *Bemesia tabaci* (QiuBaoli *et al*., 2008) and sugar beet beetle (Saleh *et al*. 2009).

Collection and handling of EPN

The best method to collect EPN is by the use of trap insects (Bedding and Akhurst, 1975). The greater wax moth, *Galleria mellonella* and meal moth, *Corcyra cephalonica* arequite suitable to EPN since they are highly susceptible, easy reared and widely available.

- Collect soil sample at a depth 10-15cm from target habitats. Place 2-5 trap hosts at the bottom of a container with capacity of 1 litre jar and lightly pack a soil sample (1kg) over them.

- Incubate at appropriate temperature (23-25°C) for maximum activity of nematodes for 5-7days.
- Collect dead hosts for dissection in Ringer's solution or Saline solution or for placement in White's trap (White, 1927) to harvest emerging infective juveniles (IJ) for further studies.
- The nematodes may be handled using a pick specially designed to lift nematodes.

Ringer's solution		Saline solution	
NaCl	: 9g	NaCl	: 10g
KCl	: 0.4g	Distilled water	: 1 lit
$CaCl_2$	: 0.4g		
$NaH_2\ Co_3$	: 0.2g		
Distilled water	: 1 lit		

White's trap

- Place a 9cm Whatman No.1 filter paper in a concave side up watch glass in a large glass Petri dish (150 X 20mm).
- Autoclave for 20 minutes at 121°C.
- Pour about 70ml sdw or 0.1% formalin into the Petri dish. Do not put any water into the watch glass.
- Drape the filter paper over the watch glass so that it comes into contact with the liquid surface.
- Place infected larvae (10-30 as they fit) on the filter paper over the edge of the water glass. IJ will start to exit 8-12 days after infection.
- Once nematodes begin to come out from the insects, they should be harvested daily until production drops (3-4 days).
- To collect, remove the watch glass with dead larvae, pour the IJ into a beaker (rinse the Petri dish to collect all the IJ), add 70ml of sdw or 0.1% formalin and replace watch glass.

Storage

The nematodes may be stored in a canted tissue culture flask containing sdw with a drop of triton X-100 (wetting agent that prevents nematode from sticking to the side of the container) or 0.1% formalin. If stored without aeration IJ should be concentrated to not more than 10,000- 20,000 IJ/ml and water depth should be 1cm or less. Higher concentrations (1,00,000/ml) will not be detrimental in aerated suspension. An aquarium pump or any forced air supply can be attached to an aquarium stone (sieve) in the nematode suspension.

Steinernematids can be stored at 4-10°C for 6-12 months without much loss of activity. Heterorhabditids do not store as well and 2-4 months of storage at 4-10°C is considered good. Heterorhabditids form rosettes (clumbs) under storage. Addition of sodium bicarbonate solution (1g NaH_2Co_3 in 50ml of H_2O) breaks up the clumbs.

Processing Nematodes to Glycerin

Dust and dirt particle will adhere to specimens and make them difficult to examine, so it is best to use clean container and solutions. Filter the solutions if and when necessary.

Put fixed nematodes in an ethanol- glycerin -water solution (15parts 95% ethanol, 1 part glycerin and 5 parts water) in a small and clean dish.

Use a glass slide or cover slip to cover all but 1/8 of the surface area for 2 days and then all but 1/4 for 7 days. The alcohol and water will evaporate to leave the nematodes in pure glycerin. In areas with high relative humidity, the dish should also be left for 5 days in a desiccator.

Surface Sterilization

Immerse nematodes in Merthiolate (0.1% w/v) + streptomycin (5000 units/ml) solutions for 2-3h. Nematodes should only lightly cover the bottom of the receptacle; otherwise, IJ on the very bottom may not maintain enough exposure to antibiotic treatment. Triple rinse in sterile distilled water (sdw). To rinse, either as aseptically pipette the nematodes from one container to the surface of the next container and allow them to settle, or align a series of separating funnels containing sdw and, after the IJ settle in each funnel, turn the stopcock to allow them to drop by drop to the next funnel.

To check that the IJ are surface sterilized (they should still contain symbiont bacteria internally), they may be placed on nutrient agar (or other bacteriological media), after 2-4days verify that only bacterial colonies are growing.

Quantification of nematodes

Depending upon the need, EPN may be counted in one of two ways- direct count and dilution.

a. Direct count

When only small numbers (1-50) of IJ are required as for a bioassay, they may be counted individually into a micropipette or syringe while observing them under a dissecting microscope. The nematodes may then be transferred where required. The pipette or syringe must be checked to verify that all nematodes expulsed. They frequently stick inside and may need to be rinsed out.

b. Dilution

When larger numbers of IJ per sample (>50) are required or when the concentration in a suspension of over 200IJ/ml must be determined, dilutions must be made and counted.

Bioassay procedure

a. Petri dish assay

- Place a 9cm Whatman No. 1 filter paper in the lid of a 10cm Petridish.
- Apply the required number of IJ evenly over the filter paper in 1ml of water.
- Add any one host (*Galleria* or *Corcyra* larva) and cover with inverted Petri dish lid.
- Place the Petri dishes in plastic bag to conserve moisture. Incubate at 23-25°C.
- Record the mortality at 24h intervals. Remove and rinse dead insects. Dissect to verify the presence of nematodes.
- Probit analysis can be used to determine median lethal concentration (LC_{50}) and median lethal time (LT_{50}).
- For determination of LT_{50}, it will be necessary to record mortality several times a day.

b. Sand barrier assay

Sand- use the soil fraction that passes through a 200 mesh (per linear inch) sieve and is retained on a 400 mesh sieve.

Assay chamber- 5cm X 5cm PVC (polyvinyl chloride) cylinders. Covers to 6 cm plastic dishes make bottoms and lids for the cylinders.

- Put bottom on a tube and fit it with moist sand obtained by mixing 70 ml water with 1kg dried sterile sand.
- Place one host at bottom of the tube. To avoid crushing the host, first make a small depression in the sand and replace bottom.
- Apply the required number of IJ in 0.5ml of water to the top of the tube (end opposite to host).

- Place the Petri dishes in plastic bag to conserve moisture. Incubate at 23-25°C.
- Record the mortality at 24h intervals. Remove and rinse dead insects. Dissect to verify the presence of nematodes.
- Probit analysis can be used to determine median lethal concentration (LC_{50}) and median lethal time (LT_{50}). For determination of LT_{50}, it will be necessary to record mortality several times a day.

References

Bedding, R.A. and Akhrust, R.J. 1975. A simple techniques for the detection of insect parasitic rhabditid nematodes in soil. *Nematologica*, 21: 109-110.

Gaugler, R. (ed.) 2002. *Entomopathogenic Nematology*. CAB International, Wallingford, UK.

Kaya, H.K. 1985. Entomogenous nematodes for insect control in IPM systems. *In*: M.A. Hoy and D.C. Herzog (eds.) *Biological control in agricultural IPM systems*. New York: Academic Press, Inc.

Subramaninan, S., Saravanapriya, B., Abirami, R. and Priya, P. 2005. A Practical guide for Entomopathogenic nematodes. Sri Sakthi Press, p18.

White, G.F. 1927. A method for obtaining infective nematode larvae from culture. *Science*, 66: 302-303.

23

Field Potential of Entomopathogenic Nematodes Against Different Economically Important Insects

C. Sankaranarayanan

ICAR-Sugarcane Breeding Institute, Coimbatore - 641 007, Tamil Nadu

Introduction

Agricultural production in India is constrained by number of biotic and abiotic factors. India is a tropical country and has a climate which is conducive for insect pests which are, one of the major constraints in agricultural production, processing and storage. Annual crop losses due to insect pests and diseases in India are estimated to be 18 percent of the agricultural output. Insect pests are widespread in all the agroclimatic zones and large quantities of insecticides are used to control these pests. Indiscriminate use of insecticides, lack of adequate knowledge on proper application techniques and dosages, and frequent use of broad spectrum insecticides are common in India. There is a need to identify suitable alternative methods for the management of insect pests and among them, biological control using viruses, bacteria, fungi and entomopathogenic nematodes is gaining greater attention world over.

Nematode families of Steinernematidae and Heterorhabditidae attracted most attention as they contain the entomopathogenic nematodes (EPN). Entomopathogenic nematodes are obligate parasites of insects that kill their hosts with the aid of bacteria (*Xenorhabdus* and *Photorhabdus*) carried in the nematode's alimentary canal. The lethal obligate insect parasites are ubiquitously distributed without pathogenicity to mammals and beneficial non-target insects, and exempted from registration requirements. Currently there are about 98 valid EPN species belonging to *Steinernema* (79) and *Heterorhabditis* (19). Soil surveys conducted in divergent areas of Indian subcontinent have demonstrated high diversity of EPN across seasons, habitats, and geographic regions. There are several target pests in India that can be controlled with entomopathogenic nematodes and several workers have used EPN against cutworms, stem borer, white grubs, etc. under laboratory and field conditions and are compatible with existing IPM programmes.

Bio-efficiency of EPN Against Insect Pests

In India, utilization of entomopathogenic nematodes started during 1966. Rao and Manjunath (1966) made first attempt to evaluate DD-136 against insect pess of rice, sugarcane and apple under field condition. In the 1970s Singh and Bardhan (1974) and Singh (1977) worked on mortality of cutworms, life cycle, filed efficacy of DD-136 and its compatibility with insecticides and fertilizers. High mortality of cutworm, *Pseudalatia separata* and *Cirpis compta* was obtained with *S. carpocapsae* (Isreal *et al.,* 1969). Work on heterorhabditids is strated during 1990s. Since 1990s, there was a great progress in EPNs research work initiated from different parts of the country and the following laboratories were actively involved in various aspects of EPN research; ICAR-Sugarcane Breeding Institute, Coimbatore; TNAU, Coimbatore; ICAR-NBAIR, Bengaluru; GAU-Anand; ICAR-IARI, New Delhi; ICAR-CPCRI, Kayankulam; ICAR-DRR, Hyderabad; ICAR-CICR, Nagpur; and RAU-Udaipur.

Sugarcane Pests

A new species of entomopathogenic nematode (*Heterorhabditis* sp. nov.) was isolated from sugarcane top borer, *Scirpophaga excerptalis* (Pyralidae: Lepidoptera) in the vicinity of Coimbatore, India (Anon, 1991) and then it was described by Poinar *et al.* (1992) as *Heterorhabditis indicus* sp. nov. first ever EPN described from India. Among the various species of white grubs infesting sugarcane in India, *Holotrichia serrata* and *Leucopholis lepidoptera* are widely distributed. Much work has been done at ICAR-Sugarcane Breeding Institute, Coimbatore to control white grubs with EPN. Survey was conducted in white grub endemic areas of sugarcane to collect EPN specific to white grubs and In two field trials *H. indica* (isolate DSM 78) caused 30 to 78 per cent reduction of white grub population.

Rice Pests

Israel *et al.* (1969) observed high mortality of paddy cutworm (*Pseudoletia separata*) and leaf folder caterpillar (*Cirphis compta*) when inoculated with the DD 136 strain (*S. carpocapsae*) in laboratory, glasshouse and field experiments. Other important rice pests such as paddy stem borer (*Tryporyza incertulus*), Asian rice borer (*Chilo suppressalis*), ragi pink borer (*Sesamia *inferens*), paddy butterfly (*Melanitis ismene*), rice skipper (*Paranara mathias*), rice leaf folder (*Cnaphalocrocis medinalis*) and gall midge (*Pachydiplosis oryzae*) were also highly susceptible. Srinivas and Prasad (1991) reported 98% mortality of the fifth instar larva of the rice leaf roller by *S. carpocapsae*. Addition of 2% glycerine and agar solutions to the nematode spray suspension is effective in enhancing the efficacy of nematodes (Yadava and Rao, 1970). The DD 136 strain is tolerant to concentrations of fertilizers and some common insecticides that prevail in rice fields (Rao *et al.* 1975).

Other Crop Pests

Singh (1977) reported 100 per cent mortality of *Agrotis ipsilon, A. segetum* and 75% mortality of white grubs due to *Steinernema* sp. The DD 136 strain can cause 100% mortality of potato chafer grub (*Anomala* sp.) (Rajeswari Sundarababu *et al.,* 1984). Three strains of *H. bacteriophora* and DD 136 were tested by Bhaskaran *et al.* (1994) on red hairy caterpillar, *Amsacta albistriga* Wlk. in field. DD 136 was more effective. The LC_{50} values ranged between 6.5 and 19.4 IJs/larva. Sezhian *et al.* (1996) enhanced the effectiveness of *S. carpocapsae* by adding glycerine 0.1% and Triton X 100 against fourth instar of *S. litura* on sunflower heads. Patel and Vyas (1995) observed 24.5% mortality of *H. armigera* on chickpea due to *S. glaseri.*

Field efficacy of EPN utilised against various pests in different insects are presented in Table1.

Table1. Field efficacy of EPN utilised against different insect pests

Pest	EPN	Application rate	Lab /field	Effect	References
Maize					
Agrotis ipsilon	*H. indica* + *M. anisopliae*		Field	Higher yield observed	Bhagat *et al.* (2008)
Cabbage & Cauliflower					
Plutella xylostella	*S. carpocapsae* *H. indica* *S. carpocapsae*	2 billion /ha	Field	52.3 % control of larvae	Saurav Gupta *et al.* (2011)
Cardamom, Tea, Arecanut & Ginger					
Conogethes punctiferalis	*S. glaseri*		Lab and Field	Mortality of shoot borer larvae was observed within 24-28 h	Balu and Varadarsan (1991)
Leucopholis lepidophora	*H. indica*	20 lakhs juveniles /palm	Field	44 – 55% mortality of grubs in arecanut	Murthy *et al.* (2010)
Laspeyresia hipunctata	*H. indica*	2 X10^9/ha	Field	Tea yield increased and effective control of insect	Devrajan *et al.* (2010)
Chickpea					
Helicoverpa armigera	*S. riobrave* *Heterorhabditis sp.*	50,000 IJs/M^2	Field	37 - 49%, reduction in pod damage	Vyas *et al.* (2003)
Helicoverpa armigera	*S. carpocapsae*	600 IJs/plant	Field	41.67% mortality of larave	Yadav *et al.* (2008)
Helicoverpa armigera	*S. masoodi*	3×10^9 IJs/ha	Field	Lowest pod damage recorded and was 68.6% increase in yield.	Ahmad *et al.* (2009)
Helicoverpa armigera	*S. seemae*	1×10^9 IJs/ha	Field	Lowest pod damage recorded and was 63.05% increase in yield.	Ali and Asif (2011)
Helicoverpa armigera	*S. masoodi, S. carpocapsae H. indica*	3 to 5 × 10^9 IJs/ha	Lab and Field	Reduced pod damage up to 11 to 12 %	Abid *et al.* (2014)

Groundnut					
Amsacta albistriga	*S. carpocapsae* *H. bacteriophora*		Field	*S. carpocapsae* was the most effective	Bhaskaran *et al.* (1994)
Potato					
Anamola	*Neoaplectana carpocapsae*		Lab and Field	Infection and EPN development observed	Rajeswari Sundarababu *et al.* (1984)
Agrotis ipsilon	*S. riobrave*	100000IJs /m^2	Field	Suppression of larval population	Mathasoliya *et al.* (2004)
Brahmina coriacaea	*H. indica*		Lab and Field	Reduction in grub population observed	Chandel *et al.* (2005)
Brahmina coriacea	*H. indica* *S. carpocapsae*	1, 3 and 6×105 IJs/M^2	Lab and Field	Reduction of grub population up to 80%	Anupam Sharma *et al.* (2009)
Brinjal					
Myllocerus discolor	*H. indica S. glaseri*	10000IJs/M^2	Field	92 – 93% reduction of gru population	Prabhuraj *et al.* (2000)
Leucinodes orbonalis	*S. carpocapsae*		Field	Reduction in borer incidence	Hussaini *et al.* (2002a)
Sugarcane					
Holotrichia serrata	*H. indica*		Field	Reduction of grubs	Mehta (1998)
Holotrichia serrata	*H. indica*	2-5 x10^9 IJs/ha	Lab, Potculture, microplot and Field	30 to 78% white grub population reduction	Sankaranarayanan and Singaravelu (2013)

Application Technology of EPN

Entomopathogenic nematodes can be applied with nearly all agronomic equipments including pressurized sprayers, mist blowers and electrostatic sprayers (Shapiro-Ilan *et al.,* 2006). Entomopathogenic nematodes should be applied at the first sign that a pest population is initializing to cause damage. Reapplying nematodes depends on the success of the first nematodes released. Their survivorship and success are based on environmental condition, moisture and soil type and percentage of living nematodes actually released during the first application. Nematodes should be reapplied on seven day intervals if damage continues. In order to ensure maximum effectiveness, it is crucial to apply them at the optimum environmental conditions needed for their better survival. Therefore, it is best to irrigate the target site, both before and after application because they need moist conditions to prevent desiccation and aid with movement to find hosts. Also, the best results are obtained when the relative humidity is high, ambient temperature is neither extremely hot or cold, soil temperature is between 10 to 35°C, soil is moist and direct sunlight is minimal. All of these factors help prevent the nematodes from drying out and increase their survival and virulence. The paramount importance, to be effective, EPNs usually must be applied to soil at minimum rates of 2.5 x 109 IJs/ha (=25/cm^2) or higher (Shapiro-Ilan *et al.,* 2002). Recycling potential should also be considered. Generally, as long as environmental conditions are conducive, nematode populations will remain high enough to provide effective pest control for 2 to 8 weeks after application (Kaya, 1990).

Acknowledgement

The author is grateful to Dr. Bakshi Ram, Director, ICAR-Sugarcane Breeding Institute, Coimbatore for the encouragement and facilities provided.

References

Abid, H.M., Prasad, C.S., Razi Ahmad, Wasim Ahmad and Rajendra Singh. 2014. Foliar application of entomopathogenic nematodes, *Steinernema masoodi, S. carpocapsae* and *Heterorhabditis indica* for management of legume pod borer, *Helicoverpa armigera* infesting chickpea. *Vegetos*, 27: 195-199.

Ahmad, R., Ali, S.S. and Rashid Pervez. 2009. Field efficacy of *Steinernema masoodi* based biopesticide against *Helicoverpa armigera* (Hubner) infesting chickpea. *Trends in Biosciences,* 1: 23-24.

Ali, S.S. and Asif, M. 2011. Comparative field efficacy of dust formulation and liquid formulation of *Steinernema seemae* based biopesticide and other IPM options against *Helicoverpa armigera* (Hubner) infesting chickpea. *Trends in Biosciences*, 4: 35-37.

Anon, 2000. *Annual Report*. Sugarcane Breeding Institute, Coimbatore.

Anupam Sharma, Thakur, D.R. and Chandla, V.K. 2009. Use of *Steinernema* and *Heterorhabditis* nematodes for control of white grubs, *Brahmina coriacea* Hope (Coleoptera: Scarabaeidae) in potato crop. *Potato Journal* 36: 160-165.

Balu, A. and Varadarasan, S. 1991. Evaluation of entomogenous nematode *Steinernema* spp. on insect pests of cardamom. In: Symposium held on June 27-28, 1991 at Kottayam, Kerala, India, pp9.

Chandel, R.S., Chandla, V.K. and Dhiman, K.R. 2005. Vulnerability of potato white grub to entomogenous fungi and nematodes. *Potato Journal,* 32: 193-194.

Devrajan, K., Subramanian, S. and Prabhu, S. 2010. Biological control of tea flush worm, *Laspeyresia bipunctata* using an entomopathogenic nematode, *Heterorhabditis indica. Indian Journal of Nematology,* 40: 145-147.

Ganguly, S. 2006. Recent taxonomic status of entomopathogenic nematodes: a review. *Indian Journal of Nematology,* 36: 158-176.

Isreal, P., Rao, Y.R. V.J., Prakash Rao, P.S. and Verma, A. 1969. Control of paddy cutworms by DD-136, a parasitic nematode, *Current Science* 16, 390-391.Symposium on '*Rational approaches in Nematode Management for Sustainable Agriculture'*, GAU, Anand, Gujarat, India, November 23-25, pp 19.

Jothi, B.D. and Mehta, U.K. 2007. Impact of different temperatures on the infectivity and productivity of entomopathogenic nematodes on *Galleria mellonella. International Journal of Nematology,* 17: 158-162.

Karthikeyan, K. and Sosamma J. 2009. Bioefficacy of white muscardine fungus, *Beauveria bassiana* (Balsamo) Vuillemin and entomopathogenic nematode, *Heterorhabditis indica* (Poinar) against rice blue beetle, *Leptispa pygmaea* Baly. *Journal of Biological Control,* 23: 79-81.

Kaya, H.K. (1990). Soil Ecology. In: Entomopathogenic nematodes. *Annual Review of Entomology,* 38: 181-206.

Mathasoliya, J.B., Maghodia, A.B. and Vyas, R.V. 2004. Efficacy of *Steinernema riobrave* against *Agrotis ipsilon* Hufnagel (Lepidoptera: Noctuidae) on potato. *Indian Journal of Nematology*, 34: 177-179.

Poinar, G.O.Jr,. Karunakar G.K. and David, H. 1992. *Heterorhabditis indicus* n.sp. (Rhabditida : nematoda) from India: Separation of *Heterorhabditis* spp. by infective juveniles. *Fundamental and Applied Nematology,* 15: 467-472.

Prabhuraj, A.,Viraktamath, C.A. and Kumar, A.R.V. 2000. Field evaluation of entomopathogenic nematodes against brinjal root weevil, *Myllocerus discolor* Boh. *Pest Management in Horticultural Ecosystems*, 2: 149-151.

Rao, V.P. and Manjunath, T.M. 1966. DD-136 nematodes that can kill many insect pests. *Indian Farming,* 16: 143-44.

Sankaranarayanan, C. and Singaravelu, B. 2012. Effect of entomopathogenic nematodes and entomopathogenic fungi against white grub *Holotrichia serrata* F. (coleoptera: scarabaeidae). Abstract of paper presented in the *International Symposium on New Paradigms in Sugarcane Research* (ISNPSR-2012) on 15-18, October, 2012 at Coimbatore. pp. 267.

Sankaranarayanan, C. and Singaravelu, B. 2013. Isolation of entomopathogenic nematodes from white grub endemic areas of sugarcane and its biocontrol efficacy against white grub *Holotrichia serrata* under field conditions. Abstracts of paper presented in National Nematology symposium on *"Nematode: A Friend and Foe of Agri-Horticultural Crops"* held at Solan from 21st to 23rd November 2013 pp. 145.

24

Field Application of Entomopathogenic Nematodes, Formulations and Future Prospects

J. Gulsar Banu

ICAR-Central Institute for Cotton Research, Regional Station Coimbatore – 641 003, Tamil Nadu

Entomopathogenic nematodes (EPN) in the family Steinernematidae and Heterorhabditidae have enormous potential to control insect pests of economic importance. Since, chemical pesticides pose environmental hazards, these nematodes have received considerable attention as potential bio-insecticides because of their wide host spectrum, active host seeking, killing the host within 48 h, easy mass production, storage and application and are environmentally safe. One of the important criteria for a successful biocontrol agent is formulation of nematodes into a stable product, which has played a significant role in commercialization of these biological control agents. Apart from formulation field application and desired control of insect pests decides the success of EPN.

Formulation of Entomopathogenic Nematodes

EPNs have been known since 1929 but they became commercially available only during 1990s.The first attempts at formulating EPN were initiated in 1979 but the shelf life was very limited (one month). Infective juveniles carried on moist substrates such as sponge, vermiculite and peat require continuous refrigeration to maintain their viability. While formulating nematodes, two things are necessary, a long shelf-life and easy application method. The simplest method is impregnating a moist substrate (such as sponge, peat, vermiculite, etc.) with nematodes. The sponge needs to be squeezed in water before application, to release the nematodes, whereas nematodes from other carriers could be applied directly to the soil. However, these formulations are labour-intensive, require continuous refrigeration and lack economy of scale and therefore can be used on small scale only.

Although the infective juveniles of entomopathogenic nematodes can be stored for several months in water in refrigerated bubbled tank, high cost and difficulties of maintaining quality preclude the deployment of this method. Therefore, nematodes are usually formulated into solid or semi-liquid substrates soon after they are produced.

Formulations with Actively Moving Nematodes

Nematode placement on or in inert carriers provide a convenient means to store and ship small quantities. Though inert carrier formulations are easy and less expensive to make but not effective under field condition.

Sponge

Georgis (1990) made the sponge formulation placing Steinernematids and Heterorhabditids onto clean, polyether-polyurethane sponges at rates of 50-100 nematodes per cm^2 surface area. An aqueous suspension was prepared by hand - squeezing the sponge in a small volume of water to extract nematodes from the sponge. Nematodes on sponges could be stored for 1-3 months at 5-10°C. About 5 - 25 x 10^6 infective juveniles were placed on a sheet of sponge that was then placed in a plastic bag. The bags were placed on ice packs for shipping. *H. bacteriophora* and *H. marelatus* in sponge formulations were found effective against *Otiorhynchus sulcatus, Popillia japonica* and *Chrysoteuchia topiara.* Owing to the time consuming and physically demanding removal method, this formulation is not suitable.

Vermiculite

The vermiculite formulation was a significant improvement over sponge formulation. An aqueous nematode suspension was mixed homogeneously with vermiculite and the mixture was placed in thin polyethylene bags. The vermiculite - nematode mixture was added to the spray tank directly, mixed in water and applied as spray or drench.

Formulations with Reduced Mobility Nematodes

Since the nematode activity was high in inert carriers and the stored energy reserves of infective juveniles were found to be rapidly depleted. Hence formulations have been developed in which the mobility of nematodes is minimized either through physical trapping or by using metabolic inhibitors.

Physical Trapping

Alginate Gel

Entomopathogenic nematodes are trapped in thin sheets of calcium alginate spread over plastic screens. Nematodes were released from alginate gel matrix by dissolving it in water with the aid of sodium citrate. The calcium alginate formulation increases the shelf-life of *S. carpocapsae* up to 3-4 months. However, the disposal of thin sheets of plastics and time-consuming extraction steps were some of the problems which rendered this formulation unsuitable for commercial application.

Flowable Gels

A formulation in which nematodes were mixed in a viscous flowable gel or paste to reduce activity was prepared. However, nematode shelf life was shorter than in alginate gels. A paste formulation was prepared in which nematodes were mixed in a hydrogenated oil and acrylamide. The nematodes could be easily entrapped in calcium alginate matrix by a fast, gentle, aqueous, room temperature process. *S. feltiae* in the capsules survived for eight months with no detectable loss in infectivity when stored at 4°C.

Metabolic Arrest

In this formulation, nematodes were trapped in powdered activated charcoal that served as an adsorbent. The charcoal and nematode mixture was stored in sealed containers. Addition of a proprietary metabolic inhibitor reduces oxygen demand and enables storage of concentrated *S. carpocapsae, S. feltiae* and *S. riobrave* without bubbling air for extended periods at room temperature. Over 7 x 10^9 infective juveniles of *S. carpocapsae* could be stored for six days at room temperature in a 10 lit. container without loss in viability. This liquid concentrate formulation is used for shipping *S. riobrave* for application on citrus against *Diaprepes* root weevil.

Formulations with Anhydrobiotic Nematodes

Here induction of partial anhydrobiosis in nematodes had been achieved successfully by controlling water activity of the formulations.

Bait

Bait formulations of entomopathogenic nematodes was tested on black cut worms, *Agrotis ipsilon* and tawny mole crickets, *Scapteriscus vicinus* by adding hyroxyethyl cellulose and glycerine to nematodes to minimize their migration from the bait. However, the bait formulations did not outperform the aqueous nematode suspension when applied against black cut worms, although it produced a significant pest reduction compared to control. When trap stations were used that ensured nematode contact with the target pest and protected nematodes from sunlight and desiccation, the baits outperformed the standard chemical insecticides against adult house flies in pig units and German cockroaches in apartments.

Gels

Steinernematids and Heterorhabditids are compatible with many gel-forming polyacrylamides. Nematodes were placed on the gel and packed inside a mesh bag from which they are easily extracted in water. The nematodes were partially desiccated, but survival at room temperature was low.

Powders

Here nematodes were mixed in clay to remove excess surface moisture and to produce partial desiccation. The formulation, termed as 'sandwich', consisted of a layer of nematodes between two layers of clay. This formulation was commercialized in Australia but was later discontinued due to inconsistent storage stability and clogging of spray nozzles. Later an improved wettable powder formulation was developed which enabled storage of Heterorhabditids and Steinernematids at room temperature (2.0 – 3.5 months).

Granules

In granular formulations, nematodes were partially encapsulated in lucerne meal and wheat flour. Later an extruded or formed granule in which nematodes were distributed throughout a wheat gluten matrix. This 'Pesta' formulation included a filler and a humectant to enhance nematode survival. However granules rapidly dry out during storage resulting in poor nematode survival.

A water-dispersible granular (WG) formulation was developed in which infective juveniles were encased in 10-20 cm diameter granules consisting of mixtures of various types of silica, clay, cellulose, lignin and starch. The granular matrix allowed access of oxygen to nematodes during storage and shipping. At optimum temperature, the nematodes entered into a partial anhydrobiotic state owing to the slow removal of their body water by the formulation. The improvement in shelf-life was attributed to the induction of partial anhydrobiosis by a mixture of ingredients in the formulation such as silica, clays, cellulose, lignin and starch.

Desiccated Cadavers

Nematode-infected cadavers have also been used, but these formulations lack economy of scale since they involve *in vivo* production and there are difficulties in fixing dose, storage and application. This is the first commercial formulation that enabled storage of *S. carpocapsae* for over six months at 25°C. Nematodes applied in the form of infected wax moth larvae were found to be as effective as aqueous nematode suspension against soil pests. It enhanced desiccation tolerance, adhered effectively to the cadaver and did not negatively affect nematode reproduction or infectivity.

In India the formulation developed by Multiplex Biotech Pvt. Ltd., Bangalore is comprised of hydro gel based semisolid cream with IJs suspension containing 1 million nematodes in a polythene pouch that can be readily mixed in a water spray tank without blocking the nozzles. The nematode shelf life of 2-3 months was achieved at room temperature (27-28°C) with maximum of 80% survival by increasing the nematodes metabolic rate mixing silica powder (0.1%). NemaGel, a formulation based on new hydrogel, has been developed at IARI for *S. thermophilum* and is reported to have storage life of 9 months.

Expected storability of some commercially available formulations of entomopathogenic nematodes

Formulation	Nematode Species	Storage duration	
		Room Temperature	Refrigerated
Liquid concentrate	*S. carpocapsae*	5–6 days	12–15 days
	S. riobrave	3–4 days	7–9 days
Sponge	*S. carpocapsae*	0 days	2–3 months
	H. bacteriophora	0 days	1–2 months
Vermiculite	*S. feltiae*	0 days	4–5 months
	H. megidis	0 days	2–3 months
Alginate gels	*S. carpocapsae*	3–4 months	6–9 months
	S. feltiae	0.5–1 month	4–5 months
Flowable gels	*S. carpocapsae*	1-1.5 months	3–5 months
	S. glaseri	0 days	1–1.5 months
Water dispersible granules	*S. carpocapsae*	4–5 months	9–12 months
	S. feltiae	1.5–2 months	5–7 months
	S. riobrave	2–3 months	4–5 months
Wettable powder	*S. carpocapsae*	2.5–3.5 months	6–8 months
	S. feltiae	2–3 months	5–6 months
	H.megidis	2–3 months	4–5 months
Nematode wool	*H. bacteriophora*	21 days	unknown

Field Application of Entomopathogenic Nematodes

Entomopathogenic nematodes can be applied with nearly all agronomic or horticultural ground equipment including pressurized sprayers, mist blowers, and electrostatic sprayers or as aerial sprays. The application equipment used depends on the cropping system. It is important to ensure adequate agitation during application.

Biotic Factors Affecting Application Success

A number of factors related to the nematode are critical for application success. The appropriate nematode must be matched with the particular target pest. Factors that must be considered in choosing the appropriate nematode include virulence, host finding, and environmental tolerance and in some cases persistence. In order to be effective, EPNs usually must be applied to soil at minimum rates of 2.5 x 10^9 IJs/ha (=25/cm^2) or higher. Depending on the target pests, a higher rate of application may be required (or in some rare cases lower rates may suffice). As long as environmental conditions are conducive, nematode populations will remain high enough to provide effective pest control for 2 to 8 weeks after application. All of the formulations are intended to be mixed with water to release the nematodes through common application equipment such as small pressurized sprayers, mist blowers, electrostatic sprayers, or even helicopters (aerial application). One of the more promising methods for applying entomopathogenic nematodes uses irrigation systems in a manner similar to chemigation.

Abiotic Factors Affecting Application Success

Successful application of EPNs depends on several critical factors including protection from ultraviolet radiation, adequate soil moisture/relative humidity, and temperature. EPN applications for above ground pests have been severely limited due to environmental hindrances (e.g., UV radiation or desiccation) that reduce survival and efficacy. So biocontrol success is most likely achieved when EPNs are applied to soil or cryptic habitats. Because ultraviolet radiation is detrimental to nematodes, applications are best applied in the evening or early morning hours, or exposure to ultraviolet radiation avoided, through subsurface application. For soil applications, moisture for EPN survival and movement is required, but too much moisture may cause oxygen deprivation and restrict movement. Thus, irrigation is recommended for maintaining adequate moisture. Optimum moisture levels will vary by nematode species and soil type. Similarly, Optimum temperatures for infection and reproduction will also vary among nematode species and strains.

Soil parameters can also be important for surface or below-ground applications. Soil texture affects nematode movement and survival. Generally, compared with lighter soils, soils with higher clay content restrict nematode movement and have potential for reduced aeration, which, in combination, can result in reduced nematode survival and efficacy. Soil pH can affect natural EPN distributions. A soil pH of 10 or higher is likely to be detrimental to EPN applications, whereas a range of 4-8, is not likely to have any significant effect on EPNs.

Important Points to be Considered while Applying EPN for Pest Management

- Nematodes require a film of water around soil particles to move through the soil profile in search of a host. Therefore, pre-irrigate the soil in the treatment area with about 0.25 to 0.5 inch of water no later than a few hours before application of the nematodes. Further irrigation to maintain adequate soil moisture for at least 7 days following nematode application also is recommended.
- Excess water inhibits the movement of oxygen in the soil, and the nematodes will drown.
- When applying agrichemicals in the area where entomopathogenic nematodes are to be used, there should be enough separation time between applications of toxic compounds and entomopathogenic nematodes.
- Soil temperature where nematodes are to be applied should be above 55°F and less than 90°F.
- Protect nematodes from excessive exposure to ultra violet (UV) rays which can inactivate and kill them.
- Time of application of entomopathogenic nematodes to target the susceptible stage of the pest. Select the proper nematode species to match the most susceptible pest stage.
- Select the application rate and method to maximize contact between entomopathogenic nematodes and the target pest.
- Proper application method and timing are crucial to success of any biopesticides because application process delivers the infective stage of bioagent to the target insect.
- Quality of commercial nematode products is critical for acceptance of EPN as biopesticides. This is very important as contamination with free-living nematodes at any stage can ruin the final product. This can be ensured by accrediting some laboratory which can be authorized to test the commercial formulations and authenticate/certify them.
- Inundative high concentration of specific species is used as a primary control strategy, mostly targeted to the soil and cryptic habitats.

Future Thrusts

- Identification of new strains and species that are superior to currently commercialized nematodes that can rapidly result in enhanced efficacy.

- Use of improved strains with enhanced levels of various beneficial traits such as environmental tolerance, virulence, reproductive capacity, etc.
- Development of more stable formulations that are simpler to use and have longer than one year shelf life.
- New nematode isolates should be collected to increase the genetic base of material available for the creation of new products and control programmes.
- Advanced research on nematode physiology and biochemistry is necessary to help increase the predictable shelf-life, the development of more novel nematode applications, the integration of nematodes into existing pest control programs.
- Promote cooperative international projects involving scientists could provide training for Indian scientists and students in university programs.
- The technology should be such which can be passed on local entrepreneurs to produce EPN for local needs.

References

Berry, R. 2007. Application and evaluation of entomopathogens for eontrol of eest insects in Mint. In Field Manual of Techniques in Invertebrate Pathology: Application and Evaluation of Pathogens for Control of Insects and Other Invertebrate Pests. 2nd Edition by L.A. Lacey and H.K. Kaya, pp. 599-608. Dordrecht, The Netherlands: Springer.

Shapiro-Ilan, D. I., Han, R., and Dolinski, C. 2012. Entomopathogenic nematode production and application technology. *Journal of Nematology*, 44:206–217.

Shapiro-Ilan, D. I., Morales-Ramos, J. A., Rojas, M. G. and Tedders, W. L. 2010. Effects of a novel entomopathogenic nematode-infected host formulation on cadaver integrity, nematode yield, and suppression of *Diaprepes abbreviates* and *Aethina tumida*. *Journal of Invertebrate Pathology,* 103: 103–108.

Zeitfracht Medien GmbH
Ferdinand-Jühlke-Straße 7
99095 Erfurt, Deutschland
produktsicherheit@kolibri360.de